Urban Public Space

Oleg Pachenkov (ed.)

Urban Public Space
Facing the Challenges of Mobility
and Aestheticization

Bibliographic Information published by the Deutsche Nationalbibliothek
The Deutsche Nationalbibliothek lists this publication in the Deutsche Nationalbibliografie; detailed bibliographic data is available in the internet at http://dnb.d-nb.de.

Cover illustration:
Photograph by Lilia Voronkova,
Street in Rio de Janeiro, Brazil
23rd of April, 2009

Library of Congress Cataloging-in-Publication Data

Urban public space : facing the challenges of mobility and aestheticization / Oleg Pachenkov (Ed.).
　pages cm
　Includes bibliographical references.
　ISBN 978-3-631-60341-3
　1. Public spaces. 2. Public art spaces. 3. City and town life.
　4. Urban transportation. 5. Sociology, Urban. I. Pachenkov, Oleg.
　HT185.U69 2013
　307.76—dc23
　　　　　　　　　　　　　　　　　　　　　　　　　　2013012573

ISBN 978-3-631-60341-3
© Peter Lang GmbH
Internationaler Verlag der Wissenschaften
Frankfurt am Main 2013
All rights reserved.
PL Academic Research is an Imprint of Peter Lang GmbH.

Peter Lang – Frankfurt am Main · Bern · Bruxelles · New York ·
Oxford · Warszawa · Wien

All parts of this publication are protected by copyright. Any utilisation outside the strict limits of the copyright law, without the permission of the publisher, is forbidden and liable to prosecution. This applies in particular to reproductions, translations, microfilming, and storage and processing in electronic retrieval systems.

www.peterlang.de

Contents

Preface and Acknowledgments 6

Introduction 13
Urban Public Space in the Age of Mobility and Aestheticization, and the
Necessity of an Interdisciplinary Approach. Oleg Pachenkov, Lilia Voronkova

Chapter 1 31
Sharing Space with Strangers in Moving Public Places: Social Mixing and
Secessionism in Mobility. Giulio Mattioli

Chapter 2 61
Architectural Visualizations as Promoters of Urban Aestheticization.
A Visual Culture Approach. Tobias Scheidegger

Chapter 3 76
Public Art Spaces: The Dilemma of Economic Growth and Social Inclusion in
an Aesthetised Urban Context. A Case Study of Initiative "Intermediae",
Matadero Madrid, Spain. Clara Fohrbeck

Chapter 4 100
Tracing Art in Urban Public Space: the Resistive Aesthetics of Cultural Actors
in Post-Communist Romania. Laura Panait

Chapter 5 116
Contemporary Art: Between Action and Work. The Cases of Krzysztof
Wodiczko and Jenny Holzer. Celia Ghyka

Chapter 6 135
Shaping Spaces of Shared Experience: Creative Practices and Temporal
Communities. Jekaterina Lavrinec, Oksana Zaporozhets

Notes on Contributors 149

Preface and Acknowledgements

An idea of the seminar and book:
This book is based on the papers presented at a seminar of the same name organized by two persons: Oleg Pachenkov and Lilia Voronkova in 2009. It was held at Humboldt University-Berlin's Institute of European Ethnology in April 2010. The workshop emerged from our interest in the issues of urban public life and public space as well as from a shared recognition that in Russia (where both of us are from and where we do much of our work), these fields have been notably under-researched.

In 2009 when we began organizing this seminar, contemporary Russian research into urban studies had taken on a fragmented character. An adequate institutional foundation for such research was completely lacking and work in the sphere of urban studies was poorly correlated as there existed no unified professional interdisciplinary discourse. Deficiencies in empirical data were painfully felt. Interactions with 'Western' colleagues and attempts at participation in international professional conferences in the field of urban studies illustrated to us just how much Russian theoretical discourse in this sphere was lagging behind. We wanted to change the situation by holding an event where Russian scholars could start establishing a common platform, language and background for the inclusion of Russian urban studies in the global academic and transdisciplinary community. The best way to ensure the participation of Russian urban scholars in such an event was to initiate it ourselves, so that is what we did.

Certain harbingers of change within urban studies research in Russia had become manifest in recent years: the steady growth in the quantity of research, publications, dissertations, presentations, thematic compilations and academic journals, as well as the increased number of city initiatives sprouting up seemingly everywhere, all indicative of a broad change in city environment. It was not difficult to see that the demands linked with the growth and expansion of this interdisciplinary field and the practical work associated with it had not only matured but were about to reach a point where a qualitative shift will be unavoidable. We had no doubt that the next two to three years would witness a true qualitative breakthrough in the field of urban studies in Russia. We wanted to help lead that revival and so organized this seminar, to learn from and bring together our colleagues as well as to offer a space in which our German colleagues and hosts from IfEE, HU could meet their counterparts from other areas of Europe and the world, facilitating discussion around particular issues within urban studies.

The initial brainstorming for the workshop raised many themes that could address a number of the challenges urban public life and public space have been facing in the last decades: challenges of safety, of control and resistance, of appropriation and re-appropriation of urban space, of the digitalization and impersonalization of social life and many others. Following the advice of some of our German colleagues anxious about the event losing focus, we narrowed this list down and allowed two issues to remain forefront. We chose to focus this seminar on issues of mobility and aestheticization as challenges to public space in contemporary cities, and planned to keep an interdisciplinary perspective in the approach. This perspective raised a central question of the role social scientists can play in the process of producing a city's image in collaboration with other professionals: urban planners, artists, grassroots public initiatives, multiple communities, etc. The workshop was called in order to clarify these issues, and the chapters you will find contained in this book raise and sometimes even answer the deep questions that underlie them.

When thinking of the possible themes for this seminar, we found ourselves influenced by the city of Berlin itself. It impressed us with its unique urban culture and history of transformation for its city landscape as well as the appearance and content of its city life. We saw here so many examples of creative urban planning, of a variety of forms of street art and other voluntary public art that is produced and performed by citizens. Far from being proclaimed a 'public evil' to be cleaned from the face of Berlin, instead this public art seems to contribute to the construction of city identity, in making the city what it is –vivid, creative, bright and intriguing. This has not always been the dominant perspective in Berlin, as we will touch upon later in this introduction, but it made an impact on us. This context also impressed us regarding the issue of the aestheticization of Berlin, an aspect that has been already become the 'talk of the town' among urban study specialists and sociologists, particularly those who have a Marxist orientation. We saw how different urban aesthetics can be and how differently they can be used towards the aims of city development. We realized aestheticization should not automatically be reduced to gentrification and historical heritage preservation; it could take the form of public art, street art, and street performances. In Berlin we have learned that even city 'freaks' bring their own aesthetics to the city and change it.

All of these observations brought us to the theme of the seminar and this book, and we have not regretted our choice.

Structure of the book

We collected a variety of papers analyzing how issues of mobility and aestheticization are represented and handled in different ways across different world cities. Although the two notions were treated equally in the call for papers, the pool of applications that we received plainly favored aestheticization as an area of attention. This unbalance was further amplified at the chapter-submission stage. Perhaps the aestheticization scholars are more proactive in their approach to (or obsession with) their issue. Still, the theme of mobility was important to our original conception of the seminar, and, rather than edit it out, we decided to open the book with one such exploration.

In the introduction written in co-authorship by *Oleg Pachenkov* and *Lilia Voronkova*, the organizers of the seminar, the issues of mobility, aestheticization and interdisciplinarity are addressed as the main challenges that contemporary social scientists are facing when they deal with urban processes. And the cities themselves experience new processes developing along these three dimensions and changing the realities, perceptions and future of urban environment. The shift from *modernity* to the *late* or *post-modernity* is treated as one of the key factors affecting those changes and calling for the interdisciplinary approach to the understanding and transformation of cities.

The collection of chapters begins with the work of *Julio Mattioli (Chapter 1)*, who addresses the issue of public transport as mobile public space. Apart from a rich and profound overview of the literature dealing with the issue of public-ness for public and private transport in contemporary cities, Julio also offers the results of his own research, in which he attempts to "...go beyond the general division of private/public space and to analyse more in depth the practices, attitudes and experiences that are associated with private and public space in the daily lives of urban dwellers". Julio stresses the relativity and ambiguity of the 'public' nature of public transport as that of a space provoking sociability as well as indifference, tolerance as well as aversion.

Tobias Scheidegger's article *(Chapter 2)* makes a shift from the theme of mobility to that of aestheticization, though it combines both. Tobias analyses the ways that certain urban lifestyles expressed in the dress codes, gestures, and motion of abstract people are represented in architectural renderings. His interest in the renderings of not-yet-built/designed spaces and buildings is not one of an architect or designer; he approaches these images as a social scientist and treats them as a form of enacting power, a type of manipulation, not reflecting but shaping and imposing reality onto cities and their inhabitants. Tobias begins with a look at Walt Disney's sketches and ends with the current construction work going on in Zurich, and on the journey between he shows that images possess a continuity of both meanings and functions, from innocent entertainment to the means of control over spaces and people. Aesthetic manners of rendering become a tool of segregation, prescribing particular groups of people to particular urban spaces and imposing these conceptions via beautiful images onto the 'actual' realities of cities.

Clara Fohrbeck (Chapter 3) addresses the issue of the relationship between proclaimed values and performed activities. She analyses the implementation of the project of an art space in Madrid, one aimed originally at the most well-intentioned values of the inclusive approach to public art, implying a design of the space and activities openly directed at the people inhabiting the areas where urban public art projects are performed. In reality she finds the situation to be much more problematic. Is there a balance to be struck between artistic ego and egalitarianism on the one hand and the needs, attitudes and everyday realities of the dwellers of a far-from-wealthy area of a big city on the other? Everybody knows public art is an aesthetic possessing protest and democratic potential and that it can provide otherwise disenfranchised people with a voice and the means of shaping their own environment in the way they want it to look, etc. But fewer people know how to realize this in real cities with real people, some of whom are lower and working class urban dwellers, and some middle and upper middle class artists. This does not mean we should stop trying. Clara's article explores the successes and failures of this Madrid attempt, thereby providing insights into how and if true balance in this area might be possible.

Laura Panait (Chapter 4) addresses in her article a phenomenon of public art. Using the example of two art festivals that took place in two different cities in Romania (Bucharest and Cluj), she analyses the variety of forms public art can take and the different effects it can have on the life of a city. As the indicators for "testing urban tissue" she uses cultural and artistic activities performed in the public spaces of cities. Laura brings together the issues of aesthetic and politic, which she examines in the context of post-communist Romanian society. Being herself an anthropologist and an active participant in the art practices she examines, Laura chooses an interdisciplinary approach for her chapter and shifts in the text from the perspective of a social scientist to one of an artist and back, making her text rich and lively.

Celia Ghyka, another Romanian author in this collection *(Chapter 5)*, also addresses public art projects and their potential of changing urban life and space. She considers the work of two world-famous artists, Zrzysztof Wodiczko and Jenny Holzer. Her choice of focus is conditioned, on the one hand, by the fact that both artists perform their projects in urban public spaces, and on the other hand, by Wodiczko and Holzer's awareness of the notion of communication as a driving motive of public spaces. Their projects challenge the image of 'ideal' public space as potentially harmonious and suggest that we approach urban public spaces as permanently exposed to conflict and negotiation. Celia utilizes Hanna Arendt's ideas on public-ness, communication, action and work as a theoretical framework that facilitates the bringing together of these two artists, public art as such, and the notion of urban public space itself. As in Laura Panait's chapter, what follows is an examination of the role, potential, and particularities of artistic activity in the production of public space and public life of cities.

An article by *Jekaterina Lavrinec* and *Oksana Zaporozhets* finishes the series of texts *(Chapter 6)* dealing with urban public art. Oksana and Jekaterina focus on the creative actions and events that change urban scenarios, transform space, intensify urban emotions, and affect the daily lives of city dwellers. The co-authors treat their own study as "both an urban adventure and a theoretical challenge" that requires the revision of reflexive tools applied to the analysis of urban settings. The theoretical framework of this article rests upon the works of Lefebvre, Debord, De Certeau and Benjamin that Oksana and Jekaterina deftly combine with empirical observations on modern cities. By bringing together traditions of *situationism* and a "micro-optics" in observation and analysis, these scholars explore the potential of urban interventions and creative actions in the co-production of urban space and life. Besides relying on their own exciting experience of participation in situationist actions and public art projects, the authors argue on the importance of a trans-disciplinary approach in urban studies in the form of the "mutation" of researcher into "reflexive activist".

Acknowledgments

We would like to thank the Institute for European Ethnology at Humboldt University in Berlin and specifically Professor Alexa Faerber, Professor Leonore Scholtze-Irrlitz and Professor Wolfgang Kaschuba for their support of the idea and their contribution to organizing this seminar. We are also very thankful to the Alexander von Humboldt Foundation, whose support of our fellowshipsmade the seminar and also this book possible, as well as to the Georg Simmel Center at Humboldt University who financially supported the event.

We would like to express our special thanks to the art group REINIGUNGSGESELLSCHAFT (namely Henrik Meier and Martin Keil) for their contribution in designing this book and to Colleen MacDonald, Devan Aptekar, Gabriel Fueglistaler, Kristen Hendrickson and Katja Schmidt for their help with technical editing of this book.

Oleg Pachenkov, Lilia Voronkova

The seminar was organized by both of us, though Lilia Voronkova did not have enough time to invest it in editing the book. Still she was the inspiration of the entire event and deserves special thanks from me personally. Although her name does not stay among the editors, this book would never appear without her support.

Oleg Pachenkov

Introduction

Urban public space in the age of mobility and aestheticization, and the necessity of an interdisciplinary approach.

Oleg Pachenkov, Lilia Voronkova

Why public space?

It is important to begin discussions such as these with a definition of public realm and space and the relation between the two. The dominant understanding of the public realm is most often associated with the idea of citizens meeting each other in order to discuss public issues, to produce an open and free public debate, and to formulate public concern. We find such a definition of public space in the works of Hannah Arendt (1952) and Jurgen Habermas (1983; 1989), the two most influential social philosophers who formulated the idea of the public sphere.

There is another approach that tends to associate public with "sociability" – the potential of an encounter and communication between strangers. This approach is more culturally than politically concerned, and is most often associated to the names of Richard Sennett, Erving Goffman or anthropologist Clifford Geertz (Sennet, 2010). From this point of view, the public sphere is one of 'broad and largely unplanned encounter' (Scrutton, 1963, 13). It is this understanding of 'public' that is most often meant when people speak about a 'public park' or 'public place' – 'to denote a place where people who do not know each other can meet and enjoy each other's company'(Aries, 1990, 9). The issue of spatiality is therefore crucial to this approach, since it does not consider 'public' as separated from the space/place and city:

> Traditionally, this [public] place could be defined in terms of physical ground, which is why discussions of the public realm have been, again traditionally, linked to cities; the public realm could be identified by the squares, major streets, theatres, cafes, lecture hall, government assemblies, or stock exchanges where strangers would be likely to meet... The most important fact about the public realm is what happens in it. Gathering together strangers enables certain kinds of activities which cannot happen, or do not happen as well, in the intimate private realm (Sennet, 2010, 260).

One of the core characteristics of the public realm is traditionally understood in terms of its opposition to privacy (Weintraub, 1997). When we look at these most conventional definitions of the public spaces, we realize that all of them understand public space as being one of gatherings – not of mobility and "moving through". The fact of the matter is that these two key characteristics of urban public space – a gathering of citizens implying communication, and purposive publicness opposed to any form of privatization – are coming up against the challenges inherent in the realities of the late, post, or "liquid" modernity (Bauman, 2000) and there is a connection between these two notions.

Why mobility?

Hanna Arendt was concerned about the Agora and Forum or Trafalgar Square which by definition were spaces of gathering for citizens – for spending leisure time for encounters, for discussions of common interests, etc. For Jurgen Habermas, the typical public spaces were coffee and tea houses where the bourgeois gathered, read newspapers, and talked. Probably the third approach, the sociability one, is the most 'open' in this regard, and implies the possibilities of not so much gathering as co-existing in a space with the potential for interaction but also equally with the potential for the lack of it. Erving Goffman was also concerned about gatherings and interaction but provided a broad understanding of interaction that encompassed the choreography of strangers moving through a particular public space without staying there or engaging in verbal communication.

Another classical representative of this approach, Richard Sennett, defines public realm simply "as a place where strangers meet" and seems to consider *anonymity* as the main virtue of the public realm and public space (it should be mentioned that the notion of anonymity was also crucial for Hannah Arendt). Sennett pays much attention to mobility, to the boundaries and borders and their "porosity" (Sennett, 2010). But still he attributes his approach to the so-called "dramaturgical" or "performative" school on the issue of public realm and with good reason, as both theater and performance imply gathering and interaction rather than silent and ignorant movement through the space without encounter.

In these approaches, the dominant understanding of public space is *static*: it is mainly associated with the ideals of citizens meeting each other in order to discuss public issues, to form an open public debate and to formulate public concern. Even when a 'softer' or 'lighter' version of public is meant, in which public space is seen as one where strangers meet, the implication remains that people come to this space, stay there for a while, encounter one another, and use the space as a stage to perform or play particular social interactions.

On the other hand, since the mid-1970s, social scientists admit that space is less and less characterized by authenticity and more and more by increasing mobility, movement, and flow. Urry claimed mobility to be "central to the way in which people live in an increasingly 'networked society'" (2002a, 1). For some scholars it marks the end and death of the place, like Edward Relph (1976) who wrote about "placelessness", about place lacking its roots and authenticity, and about "other-directed places" full of people from elsewhere going to elsewhere. Mark Auge (1995) writes about "non-places" to mark the notions of "fleeting, the temporary and ephemeral". "Non-places" are places where traditions and authenticity are not relevant, *"marked by mobility and travel"*.

We observed this tendency in our research of city squares, as in St.-Petersburg where one of the central city squares (Sennaya) after regeneration turned from being a space of *gathering* to a space of *transit*. It used to be a place for people to meet, to stay, and to hang around. After reconstruction in 2003 it became a space to pass through, a place for very short meetings – such as picking up a friend at the metro station exit – while en route to be in another place, most often the mega mall and entertainment center newly built in the square (OPEN/CLOSED, 2010). Does this mean this square stopped being a public place while the mega mall inherited its public functions? Does this mean the square as a public space in the city could not accept the challenge of accommodating mobility as a new quality of the space? Does this mean it lost its authenticity and became instead an "other-directed" place or a "non-place"?

For other scholars this tendency towards increasing mobility marks the new meaning of place in post-modernity. In 1980s Edward Soja (1984) wrote about the underestimation of space in social theory; in the same year Doreen Massey (1984) had to convince his colleagues and readers that "geography matters". But already in 1990s the exact ways that "geography matters" to social reality were rethought. Akhil Gupta and James Fergusson (1992), as well as Appaduraj (1996), located identity and culture in mobility rather than in place. Other scholars, like Chambers, spoke about a "cosmopolitan existence" where the pleasure of travel is not only to arrive, but also *not to be* in any particular place (Chambers, 1990, pp. 57–58). Zigmunt Bauman (1998) attributes the right and possibility to move and not be tied to any particular place as a key notion of the age of globalization, as well the main privilege and powerful resource held therein. Doreen Massey (1994; 2005) in the same concern writes about the necessity for understanding a place in terms of interactions, networks and movements.

One of the consequences of increasing mobility of people is the *individualization* of society – which has seemingly supplanted privatization as the topmost threat to public spaces. For several decades critical social philosophers were defending public space from privatization. In the 1980s and 1990s the very term "privatization" became perhaps the most commonly used when addressing issues around urban public space. Since that time public has been often defined as independent from private interest and concerns, ownership, and control and the main principles of public space are considered: *1)public (not private) stewardship, 2)open access, and 3)*that the spaces are *"used by many people for common purpose"*(Zukin, 1995, p. 32, 38).

However the concept of public is now threatened not merely by privacy but by individualism (Bauman, 2000; Elias, 1991). Mobile citizens filling city spaces bring the idea of disparate molecules to mind— isolated and moving without any expressed need for gathering together, communication, or interaction; these mobile citizens see each other and a crowd as an irritating obstacle in the way of even faster movement. In this scenario, the threat to a coherent public space arises not from somebody being interested in appropriating it for private interests such as private business (though this may indeed also be the case), but rather from the blatant lack of interest in public concerns among individuals. The individual is the worst enemy of the citizen, as Alexis de Tocqueville said, so public issues are not much in demand in the society of individuals. And as a space is constituted by those activities that fulfill it and those actors that perform there, the vanishing of the *public* itself can mean the vanishing of public space. The latter is more and more often characterized by the *"void"*, by *"-less"* and *"-ness"*.

This problem is formulated by John Urry as follows:

> ... sociology, like other social sciences, has overly focused upon ongoing, more or less face-to-face social interactions between peoples and within social groups. Sociology has taken connections to be most importantly face-to-face, characterised by social interactions with those who are immediately present. However, this is problematic since there are many connections with peoples and social groupings that are not based upon regular face-to-face interactions. There are multiple forms of 'imagined presence' through diverse objects and images that carry connections across, and into, multiple kinds of social space (Urry, 2002a, p. 2)

The matter at hand concerns how we correlate these new concepts of space and place with the original notion of public space. How much and which particulars of the public remain in public space in the age of mobility? Instead of deploring the vanishing of public places, shall we re-think the very concepts that define them? Perhaps post-modernity and globalization demand new types of public places and a different understanding of the public realm and public spaces, as well as those activities that, in turn, make them into 'places'. Shall we, following John Urry (2002a; b), deny the dichotomies of mobile/immobile, presence/absence, and here/there and analyze instead the particular compositions of these aspects of social reality and their effect on the spatiality of social life? All in all, people moving in a city space are usually traveling to another place for a face-to-face meeting, i.e. their 'absence' in one particular place is aimed at them being 'present' in another one. What still changes in this construction is the 'nature' of a particular urban public space becoming transitory or a "third place", though the opposition of public/private is itself becoming ever more blurred in the mobile world (Urry, 2002a, p. 6). Besides, is public space in the city a virtue in and of itself or is it a means of performing urban public life? And when the formats of public life change rather than disappear, in which ways do public space change as well?

These many avenues point to an underlying question of whether or not to change our way of thinking. For example, look at the way that Tim Cresswell characterizes the specificity of The Hague's non-places:

While conventionally figured places demand thoughts which reflect assumed boundaries and traditions, non-places demand new mobile ways of thinking… Not only does the world appear to be more mobile but our ways of knowing the world have also become more fluid. This "weak thought" or "nomad thought" is more willing to transgress the boundaries of academic disciplines, the boundaries that separate high and popular culture and the boundaries that separate academia from the everyday world outside the ivory tower. These new kinds of thinking are symptomatic of post-modernity (Cresswell, 2003, p. 17).

contemporary cities are experiencing different forms of aestheticization

Why aestheticization?

Another global trend is the aestheticization of all aspects of social life. The story is rooted in the 17th and 18th centuries when the coherent sphere of culture, then characterized as a unified world view of religion and metaphysics, fell apart into the three separate spheres of science, morality and art. Since that time the modern world has basically been living in three separate domains, each arranged under specific aspects of validity: those of knowledge, justice and taste. "There appear the structures of cognitive-instrumental, of moral-practical and of aesthetic-expressive rationality" (Habermas, 1983, p. 9). Although all three domains were proclaimed independent and each oriented towards its own intrinsic logic and laws, science based on cognitive logic has taken primary dominance. The instrumental rationality characteristic of this domain has underlain the social order of 'Western civilization', while the moral and expressive rationalities intrinsic to the two other domains maintained a status that was legitimate but secondary.

The end of the project of modernity and the shift to postmodernity has been characterized by the growing significance of another domain – that of the aesthetic. As Pierre Bourdieu (1984) has shown, judgments of taste seem to determine the social order to an even greater extent than those of logic; this trend is gaining strength nowadays. According to a number of prominent authors "...the rapid growth of a popular culture - centered around commodities - came to play a central role in the transformation of social space itself from a cognitive to an aesthetic logic" (Desmond et al., 2001, p. 255). As Baudrillard (1993) put it, nowadays "Everything is sexual. Everything is political. Everything is aesthetic". The mode of perception has moved from words to images, and everything has to 'look like' something, have 'its face'. Moreover, it is not enough just to 'look like'– things must be designed and aestheticized. State policy and social movements, city spaces, public and private places and lives, everything around us purports to be designed and aestheticized. Even protests against aestheticization must be aestheticized.

Mike Featherstone (1991) considers the total aestheticization of social life (including everyday lives) to be a feature of postmodernity and of the urban experience particularly. He describes the contemporary city as dominated by "'de-centered subjects' who enjoy experimentation and play with fashion and the stylization of life as they stroll through the 'no place' postmodern urban spaces" (Featherstone 1991, p. 65). Therefore urban space is where mobility meets aestheticization. How does it affect public spaces?

This process is also multi-faceted; in regard to the urban landscape and the city's public culture, aestheticization can be understood in at least three different contexts:
1) the rise of cultural and creative industries and their growing role in the political economy and space production of cities;
2) the implementation of the modernist ideals of order, harmony, purity and beauty into urban planning and design;
3) the visual 'tactics of resistance' and re-appropriation of space by marginal groups and artists as realized by means of aesthetics and the aestheticized counter-culture.

The increasing role of culture and arts in cities; cultural (creative) industries.

The global aspiration to aestheticization has led to a rise in the impact of culture and arts. Since the 1970s culture and art have been playing a leading role in urban redevelopment strategies all around the world:

> Concern for "quality of place" expressed through cultural facilities and activities has certainly been a central concern of many – especially industrial – cities as part of an attempt to promote a new image to attract investment, mobile professionals, and tourists, as well as retaining existing residents (Oconnor, XinGu, 2010, p. 124).

Culture became more powerful as it became more and more part of the business of cities; along with art it has been recognized as "an engine of economic growth" for cities (Zukin, 1995, p. 110). Creative industries are replacing traditional ones and have become a significant part of the forces that produce the contemporary social, economic, and cultural space of cities. Accordingly, this has increased the role of creative industries in the production of city space itself: "The growth of cultural consumption (of art, food, fashion, music, tourism) and the industries that cater to it fuels the city's symbolic economy, its visible ability to produce both symbols and spaces" (Zukin, 1995, p. 2).

This flourishing of cultural industries has had some ambiguous consequences (Harvey, 1989; Zukin, 1991, 1995). On the one hand, "a new nexus of action houses, art galleries, art museums, art producers, and cultural and social elites – many of them international" contributed to the reputation of many cities around the world as culture capitals. On the other hand, "cultural spaces...enhance the economic value of commercial and residential property", and cultural activities transform the city spaces into "the 'clean' entertainment, commercial, and residential zones preferred by professionals, managers, and white-collar workers" (Zukin, 1995, p. 115, 118). Thus, another consequence was gentrification followed by redevelopment and 'revitalization' of the city quarters. The creative industries and the artists whose hands were deeply involved in this work were usually pushed out of these redeveloped quarters, being unable to afford the growing real estate value and rent (Zukin, 1982). All around the world artists "have been welcomed as 'bridge' gentrifiers – but not as statutory tenants deserving protection when property values rise" (Zukin, 1995, p. 111).

Another reversal of the burgeoning role of arts and culture in contemporary cities is what Sharon Zukin called an "aesthetic pressure". What this logic of aestheticization of urban space implies is a belief that the dominant form of aesthetic is the only one possible and has the universal value. This approach reduces culture to high culture and aesthetic to a very limited number of its possible forms, ignoring the fact that street art, body art, public art, etc. are also forms of culture and aesthetic.

For example, in 1990s Berlin aestheticization was expressed through the so-called "festivalization of city policy" (*Festivalisierung der Stadtpolitik*) (Becker, 1998*)* in which art and culture, festivals, and an entire urban art scene became an important part of the identity, official development, and marketing strategies of the city. But this policy turned finally to what some people named the "taste terror of one normative aesthetic" (*"Geschmacksterror einer normativen Aesthetik"*) (Becker, 1998, p. 185). The festivalization of the city has been finally reduced to a particular aesthetic that is closely linked to the tastes and ideals of certain social groups and their political preferences. Festivity in Berlin was transformed into the marketing of newly built business-centers and regenerated areas aimed at increasing the value of real estate. Aestheticization was in this manner reduced to the ideals of 'safety' and the 'emotional consolidation' of the dwellers of newly built condos, becoming a tool of gentrification. All the alternative aesthetic forms were officially condemned by the conservative city authorities and associated with disorder, dirt and danger (Becker, 1998; Holm, 2010).

Modernism and the "terror" of a particular aesthetic in urban design.

The domination and terror of a particular form of aesthetic in urban design has in fact historical roots. The public life of medieval cities was concentrated at the city markets, and was characterized by density, chaos, noise, brightness, strong smells, dirt, crowds and mixture of use. But the transition from medieval times to modernity implied the processes of regulation of spaces and activities. While the medieval city demonstrated the mixed use of public space and was characterized by bringing together different activities into the same space, the modern spatial order of the cities was rather oriented to the ideas of functional and spatial divisions among city spaces. The aestheticization of modern spaces (that began in the Renaissance but reached its peak in the 19th century) contributes to the framing of cities as cultural centers, but such an attitude implied the separation of the acceptable and unacceptable, of proper and improper, "of the 'higher' functions of cities from the chaotic, swirling mass of poverty and decay" (Zukin, 1995, p. 280). As some scholars formulate, the problem of modernist urban managers "was their inability to distinguish between disorder and diversity", they assumed that "any violation of their ideal of public order necessarily equated with anarchy" (Dennis, 2008, p. 146). These ideas and ideals were refined immaculately by le Corbusier, who

> hated the unregulated disorder of street-life; the street, he wrote, is "impure".
> ... Corbusier's intended destruction of vibrant street life was realized in suburban growth for the middle classes, with the replacement of high streets by mono-function shopping malls, by gated communities, by schools and hospitals built as isolated campuses. Over-determined pre-planning on this model has become endemic in modern urbanism: the proliferation of zoning regulations in the 20th Century is, for instance, unprecedented in the history of European urban design (Sennett, 2010, p. 263).

There were several consequences of that attitude and the domination of particular aesthetic forms in urban planning and design.

First, the modernist urban planning policy resulted in the production of empty and dead places in cities, in the expanse of space which "would have little or no connection with the other spaces of the city and could be left under-used, only to be watched from the top of the high rise buildings or from the car windows" (Madanipour, 2003, p. 202). Although modernism claimed the priority of publicness over privacy and of public spaces over private ones, the modernist obsession with the principle of order, hygiene and aestheticization led to the vanishing of public spaces across Europe. With modernism, public space seemed to have won that "permanent struggle" between public and private but some scholars consider it to have been a "pyrrhic victory". Although in the modern city, the public unconditionally dominated the private, it was a very different public space than before:

> This might appear a domination of public space, as it was drastically expended, in form of the parks in the middle of which high rise buildings were erected. However, this space was ill-defined and under-used, indeed it

was 'lost space' ..., where none of the functions of the public space could be performed; sociability was becoming impossible (Madanipour, 2003, p. 202).

The second consequence is that modernist urban design is dominated by passive spectatorship and that social functions are sacrificed to the virtues of aesthetic. This thesis is suggested and elaborated, for instance, by Richard Sennett, who shows that while before modernism the monumental buildings "were meant equally to be used" and "the spectator of the monumental building was also an actor in it", since the 19th Century major buildings in the city "came to be conceived as objects to be looked at, to be viewed" as theatrical spectacle (Sennett, 2010). Sennett calls it the "dilemma of the divide between the utilitarian and the aesthetic". The domination of this aesthetic leads to the "offer of visual pleasure at the cost of mixed social and economic use", to the "social exclusion in the name of visual pleasure" (Sennett, 2010, p. 266).

In the same regard Sharon Zukin raises the problem inherent in the preservation of historical heritage, and writes about the "curse" of aesthetic and culture in the context of urban spatiality. On the one hand, aesthetic-based historic preservation can save an area from demolition and large-scale redevelopment. On the other hand, historical and aesthetic reasons can bind an area's development (Zukin, 1995, p. 123). Moreover, the elite groups concerned about the preservation of heritage for the sake of the virtues of high culture and a particular aesthetic can "defend the architecture, or the look, or a place without thinking about the right of people to live there" – this can especially affect low-income ethnic groups (Zukin, 1995, p. 286). So historic preservation and aestheticization are never just cultural categories – there is always a political and economic one as well, and the means of performing and preserving power.

What this particular aesthetic-based tendency of history, beauty definition, and preservation underestimates is the fact that there is a multitude of possible aesthetics, and there exist social groups that prefer non-dominant ones. Although we live in an age of total aestheticization and symbolic economy, there are many different aesthetics and many co-existing systems of symbols. Trash and glamour are both aesthetics. Squat, punk, anarchy, and revolution itself all exist as forms of aesthetics too. And although there are people who want to live in clean, nice, cute, and quiet places, there are also people who prefer living in chaos, in unregulated space, in squats, and in art studios and democracy presupposes that all should have the right to perform their way of life. That is why contemporary cities are experiencing different forms of aestheticization and the "aesthetic struggle".

The visual "tactics of resistance" and the re-appropriation of space by marginal groups and artists.

If our entire culture is getting aestheticized, one can expect the same from the strategies (and tactics) of the counter-culture. The traditional form of counter-culture – a 'revolutionary' one – is moving out of time and fashion; in an age of total aestheticization, resistance needs to take different forms to be effective. Here is a list of some possible forms:

"...the resistance of advertising space (through anti-advertising and the "tactics" of resistance), resistance against roads (through tunneling and seeking to "reclaim the streets"); resistance against capitalism (through spectacular rallies); resistance against consumer society (through skip diving, No Shop Day and a myriad other actions); resistance to home ownership...; resistance around the dinner-table ...; resistance through the construction of temporary autonomous zones, including "rave" spaces and "mystic spaces" ...; the "reclamation" of the body through tattooing or piercing, resistance through the creation of web-spaces via the development of new "communities" (Gals on Web, netchicks, Napster.com)" (Desmond et al., 2001, p. 258).

We have discussed above the example of Berlin, which became a city of festivity in the 1990s, and for which festivalization became a core idea and mechanism of the official city development strategy but then transformed into the "taste terror of one normative aesthetic" formulated by conservative politicians and philistines. However, there were opposite tendencies as well. It would be a mistake to solely reduce the role of arts and festivalization in Berlin to the growth of the real estate values and gentrification. As we shared in the start of the introduction, the aesthetic of Berlin is more complex and in fact contributed to the inspiration for the focus of this seminar and book. Many aestheticized protests and public art actions took place in Berlin that were aimed at changing the direction of the development of the city architectonics, appearance, and image. What was posited in opposition to the policy of the "taste terror" and cleansing (*Räumung*) of the city was an alternative aesthetics of the squats and *Wagenburgen*, of trash and punk, of anarchists and autonoms, etc. inherited from the cultural traditions of Kreuzberg and Schoeneberg of West Berlin of the 1970s and 1980s, as well as from the Prenzlauer Berg opposition to the socialist oppressive state in late 1980s.

Therefore, aestheticization is not always and not only about gentrification. It has some positive potential and a power of resistance. A city is not only a site of power; some aspects of urban experience such as "difference, complexity and strangeness are not only forms of domination, but could also result in the resistance to domination" (Sennett, 1994, p. 26). Exhibitions, open-air performances, video and media projects, graffiti and other forms of public and street art – all these are the tools and means of the alternative aestheticization of the city public life and public space.

Yet, what is to be stressed here is the very fact that both the gentrification of the city and the resistance to it are realized by the means of aesthetic and in the aesthetic dimension. Berlin is a good example in this concern: "In Berlin the role of aesthetic and spectacle, as well as art and culture as integrative and also interventionist factor of the development of the city [in the 1990s], demonstrated themselves quite obviously" (Becker, 1998, p. 188). That is why, in spite of all the exaggerations, the slogan of Berlin (though informal) is still "Poor but sexy!" where "sexy" is more meant as an aesthetic than as a biological or ethical category.

Why interdisciplinarity?

Last but not least, we decided it would be a mistake to discuss all the above-mentioned issues solely in the narrow circles of sociologists or ethnologists. We believe that mobility and aestheticization must not be highlighted as leading principles of contemporary living unless we also include such notions as "inter-" and "cross-" – the striving for transgressing traditional disciplinary fields and boundaries. Even an interdisciplinary group of different social scientists seemed to us insufficient. We believe that terms such as "urban studies", and even more "urbanistic", imply interdisciplinarity – are understood very widely as coming over the borders of not only disciplines but also of the spheres of knowledge and systems of world views, such as science and arts, academic and practical, theoretical and applied. We believe the discussion of urban public space will benefit from the expertise of urban anthropologists, ethnographers, human and social geographers, and sociologists, as well as of urban designers and planners, artists, curators, and grassroots civil society initiatives and communities. This belief seems to reflect the trend characterized by Clifford Geertz's in the same year that Habermas wrote on the "incomplete project" of modernity:

> ... what we are seeing is not just another redrawing of the cultural map - the moving of a few disputed borders, the marking of some more picturesque mountain lakes – but an alteration of the principles of mapping. Something is happening to the way we think about the way we think (Geertz, 1998, p. 226).

We have already mentioned earlier the split of the sphere of culture into three autonomous spheres of science, morality and art, which took place during the Enlightenment (Habermas, 1983). This idea of the separation into three different domains became a dominant logic of the project of modernity. However the end or decline of modernity, which according to many scholars we are now witnessing, destroys this logic and structure. The boundaries between disciplines within the sphere of science, as well as the boundaries between the spheres of science, morality and art are becoming blurred and penetrable. Moreover, the new trend is aimed at crossing these borders. The holistic approach based on striving to understand an object under consideration in spite of disciplinary boundaries or even the ones between domains of social and cultural life is becoming a virtue.

For Klein and Newell (1998) "interdisciplinary" studies (IDS) are defined as "a process of answering a question, solving a problem, or addressing a topic that is too broad or complex to be dealt with adequately by a single discipline" (Klein and Newell, 1998, p. 3). There is already a debate in natural and social sciences, in humanities and arts about differences between "interdisciplinarity", "crossdisciplinarity" and "transdisciplinarity", and some even speak of a "postdisciplinary world":

> Multidisciplinary work draws upon knowledge from more than one discipline, but preserves the disciplinary identities of these multiple disciplinary elements. Certain objects of study – opera and the city, to give two of my favorite examples—seem naturally suited to multidisciplinary investigation. Crossdisciplinary work, in contrast, illuminates the subject of one discipline from the perspective of another, as when, for an example, a physicist discusses the acoustics of music production or a literary-studies specialist performs a "close reading" of a legal contract. In contrast to multidisciplinary and crossdisciplinary work, interdisciplinary work ideally produces knowledge that integrates two or more disciplines, contributing to a new foundational unity of understanding, perhaps even a new hybrid field. Interdisciplinary work thus both creates knowledge and redraws the boundaries of that which can, in theory, be known, but intedisciplinary work also entails an understanding of the disciplinary norms that are being challenged. /.../ The term "postdisciplinarity" evokes an intellectual universe in which we inhabit the ruins of outmoded disciplinary structures, mediating between our nostalgia for this lost unity and our excitement at the intellectual freedom its demise can offer us. Is the era of postdisciplinarity upon us now? (Buckler, 2004, p. 2).

Aestheticization of all spheres of culture and life, which became one of the focuses of this book, is only one aspect of the more overall process of "creating unconstrained interactions of the cognitive with the moral-practical and aesthetic-expressive elements" which alone, according to Jurgen Habermas, can "cure" the "reified everyday praxis" damaged by the separation of three spheres of human experience and culture (Habermas, 1983, p. 11). In fact, as Habermas claims, total aestheticization is an example of the "over-extension" of "one of these spheres into other domains" (Habermas, 1983, p. 12). Such a one-sided forcing of a single cultural sphere can cause distortions and extreme forms that are damaging to social life (Habermas gives in this regard examples of aestheticization and moralization (moral rigour)). One of the aims of our seminar and this collection of chapters was to make a contribution to the rebalancing of this process of extending (and crossing) the boundaries between science, morality, and art to benefit culture and society.

Our intention was to advance our understanding of the city (which in fact is a natural object for a multidisciplinary approach) by combining the efforts, traditions, perspectives, and advantages of the several disciplines and fields of knowledge we represent, by bringing together in this particular project science, art and morality. Therefore, representatives of a variety of disciplines and realms of activities and knowledge were very welcome at our seminar and many of them kindly accepted that invitation. The results of this attempt you are holding in your hands right now.

One more idea is worth noting in this concern is that inter/transdisciplinarity is linked not only to the notion of aestheticization; in fact all three keystones of the seminar and book – mobility, aestheticization and inter/transdisciplinarity – relate to each other at a deep cognitive level. We shall admit that crossing the disciplinary boundaries is itself a sort of 'intellectual mobility' that shares some basic principles with physical mobility: a particular state of mind assuming and accepting an idea of a move, a readiness to move and to face something new and unfamiliar, and a commitment to open one's mind and to exchange a perspective. Thus inter/cross/transdisciplinarity as an intellectual practice matches mobility both as a global trend of the late modern way of life and as an actual state of mind.

What shall we do further?

The spheres of science and art have been extensively discussed above, and here the domain of morality comes to our debate. The issues discussed in this introduction and the following chapters raise a number of questions that seem to be moral challenges for social scientists too.

Shall we – social scientists, urban planners, artists, etc. – make efforts to wake the 'citizen' inside the 'individual'? Shall we start from the classical definition of the public realm, public activities and public spaces? Should we be struggling to bring 'public' back to the life of contemporary cities and citizens?

Or do we view the time of such public life as having already passed and, instead of deploring the 'fall of public man', start thinking of something new, putting our energies into defining a new concept that will help us to better understand contemporary social reality?

Shall we deplore the very fact that the concentration of cultural industries in certain city quarters will lead to their gentrification, which would push out the artists who developed these areas (not to mention the prior non-artist inhabitants)? Or do we instead resign ourselves to this tendency as a part of the global trend of total mobility and temporality, as another example of the crossover of mobility and aestheticization?

Shall we celebrate the spaces of galleries and studios, the events of exhibitions, openings, and vernissages as new live-though-temporary spaces of public life produced by artists in the cities – instead of as markers of the vanishing traditional public spaces of squares and parks?

Shall we remember that public art is a significant "aesthetic mode of producing space" (Zukin, 1995, p. 23) and discuss the particular ways we can combine our own efforts and potentials to produce city public spaces the way *we* would like?

These and other issues are addressed by the authors of this collection of chapters.

References

APPADURAJ A. (1996) *Modernity at large.* Mineapolis: Minessota UP
ARENDT, H. (1958) *The Human Condition.* Chicago: University of Chicago Press
ARIES, P. (1990) Introduction. Aries P., Duby G. *A History of Private Life.* Vol. 3, 1987-1991. MA, Cambridge: Harvard University Press
AUGE M. (1995) *Non-Places: Introduction to an Anthropology of Supermodernity.* London & New York: Verso Books
BAUDRILLARD, J. (1993) *The transparency of evil.* Verso.
BAUMAN Z., (2000), *Liquid modernity,* Cambridge: Polity Press; Bauman Z. (2001) *The Individualized Society.* Cambridge: Polity Press.
BAUMAN Z., (1998), *Globalization: The Human Consequences.* New York: Columbia University Press.
BECKER, J. (1998) *Hype Park. Festivalisierung, Kultur-Marketing und Symbole des Widerstands in einer unternehmerischen Stadt.* In: StadtRat(Hg.) *Umkämpfte Raume.* Verlag Libertäre Assoziation. Verlag der Buchläden Schwarze Risse – Rote Strasse, p. 179–189
BOURDIEU, P. (1984). *Distinction: A social critique of the judgment of taste* (R. Nice, Trans.). Cambridge, MA: Harvard University Press.
BUCKLER J. A. (2004) *Towards a new model of general education at Harvard College* http://www.lancs.ac.uk/ias/events/general07/docs/interdisc/Interdisc-Buckler-Harvard.pdf, [Accessed: 1 December 2011]
CHAMBERS, L. (1990) *Border dialogs: Journeys in Postmodernity.* London-New York, Routledge.
CRESSWELL, T. (2003). *Introduction: theorizing place. Mobilizing place, placing mobility: the politics of representation in a globalized world.* Ed. By: GinetteVerstraete, Tim Cresswell. Thamyris intersecting place, sex and race. # 9, 2002. Editions Rodopi b.v.: Amsterdam.
DESMOND J., MCDONAGH P., O'DONOHOE S., (2001) Counter-Culture and Consumer Society. *Consumption, Markets & Culture.* Vol. 4. N 3. P. 241–280
DENNIS, R. (2008) Cities in modernity. Representations and productions of metropolitan space, 1840–1930. Cambridge University Press.
ELIAS, N. (1991) (1939) *The Society of Individuals,* Oxford: Blackwell Publishers
FEATHERSTONE, M (2007) *Consumer culture and postmodernism.* Sage Publications

GEERTZ, C. (1998) (1983) Blurred Genres: The Refiguration of Social Thought. In: William H. Newell (ed.), *Interdisciplinarity: Essays from the Literature*. New York: College Examination Board

GUPTA A., Fergusson J. (1992) Beyond culture: Space, Identities and Politics of difference. *Cutural Anthropology*: 7, 1, 1992, p. 6–22.

HABERMAS, J. (1989) [1962] *The structural transformation of the public sphere: an inquiry into a category of bourgeois society*. Cambridge Massachusetts: The MIT Press.

HABERMAS, J. (1983) Modernity – an incomplete project. In: H.Foster (ed.), *The Anti-Aesthetic: Essays on Postmodern Culture*. Seattle: Bay Press, pp. 3–16.

HARVEY, D. (1989) *The Condition of Postmodernity*. Oxford: Blackwell.

HOLM A. (2010). *Wir Bleiben Alle! Gentrifizierung – Städtische Konflikte um Aufwertung und Veränderung*. Münster: UnrastVerlag.

KLEIN, J. T. and Newell, W.H. (1998) 'Advancing Interdisciplinary Studies. In William H. Newell, ed., *Interdisciplinarity: Essays from the Literature*. New York: College Examination Board, 1998, pp. 3–22.

MADANIPOUR, A. (2003) *Public and private spaces of the city*. Routledge.

MASSEY, D. (2005) *For Space*, London: Sage

MASSEY, D. (1994) *Space, place, and gender*. Minneapolis: University of Minnesota Press.

MASSEY, D. (1984) *Spatial Divisions of Labour: Social Structures and the Geography of Production*. London and Basingstoke: Macmillan

OCONNOR, J., GU. X. (2010): "Developing a Creative Cluster in a Post-Industrial City", *The Information Society*, 26:2, 124–136.

RELPH, E. (1976) Place and placelessness. London: Pion

SCRUTTON, R. (1963) Public space and the classical vernacular. Nathan Glazer, Mark Lilla (Eds.) *The public face of architecture: civic culture and public spaces*. New York: Free Press

SENNETT, R. (1994) *Flesh and Stone: the Body and the City in Western Civilization*. New York: W.W. Norton

SENNETT, R. (2010) The public realm. Gary Bridge, Sophie Watson (Eds.) *The Blackwell City Reader*. London: Blackwell Publishers, pp. 261–272

SOJA, E. (1984*)* The spatiality of social life: towards a transformative retheorisation. In: D.Gregory & J.Urry (Eds.) *Social relations and spatial structures*. London: Macmilan, pp. 90–127.

URRY, J.(2002a) *Mobility and Connections*. Paris, April 2002, (www.ville-enmouvement.com/telechargement/040602/mobility.pdf)

URRY, J. (2002b)Mobility and Proximity. *Sociology 36,* pp. 255–74

WEINTRAUB, J. (1997) The theory and politics of the public/private distinction. In: Jeff Weintraub, Krishan Kumar (Eds.) *Public and private in thought and practice.* The University of Chicago Press, Chicago

ZUKIN, S. (1982). *Loft-living: Culture and capital in urban change.* London: John Hopkins Press Ltd.

ZUKIN, S. (1991). *Landscapes of power: From Detroit to Disney World.* Berkeley: University of California Press.

ZUKIN S. (1995) *The Cultures of Cities.* Blackwell Publishers

CHAPTER 1

Sharing Space with Strangers in Moving Public Places: Social Mixing and Secessionism in Mobility

Giulio Mattioli

Introduction: defining the terms

To address the relationship between public space and mobility is an important task but not an easy one: both notions have in fact been conceptualised in many different ways, varying deeply in breadth and level of generality. In this introductory section then, I will briefly discuss the conceptual ambiguities surrounding the two concepts in order to more clearly specify how I plan to use them in the rest of the chapter.

Much of the debate about public space has focused on its assumed erosion as a consequence of the pervasive processes of privatisation at work in many contemporary societies. However, as Sheller and Urry have pointed out, "social scientists have often not adequately distinguished" between the various meanings of 'public' and 'private' (2003, p. 108). The authors propose, then, a possible classification for the different forms of public/private boundary (pp. 109–113), which is useful to recall here:

- Private *interests* versus public interests, i.e. the boundary between the market and the state
- The public *sphere*, defined, following Habermas (1989), as "a space of rational debate and open communication mediating between the state and the private sphere of family life and economic relations"
- A private *life*, occurring within the domestic realm, set against the public life which takes place "within politics, the workplace, religion, education and other public places"
- *Public space*, defined as "those areas and locales, especially in towns and cities, outside the private spaces of the home and work, where people can congregate, socialise and organise in relatively unregulated ways"
- Mass-mediated *publicity* versus related questions of privacy[1]

---

[1] Sheller and Urry (2003) go as far as to argue that maintaining the various distinctions between public and private domains is useless and impossible, claiming that "the distinction between public and private domains should be dispensed with since nothing much of contemporary social life remains on either one side or the other of the divide"

In particular, according to Sheller and Urry, "the 'public sphere' of civil society has normally been conflated with that of 'public space'" (2003, p. 114). Similarly, Tomas (2001) has denounced "the myth of the concept of public space invented by Jürgen Habermas" (widespread in the French-speaking context for translation reasons) which has led many scholars:

> ... to relate the crisis of urban public space to an assumed crisis of collective life and democracy, to the point of confusing one concept with the other, even if that is improper and ambiguous[2]. (own translation) (p. 76).

Recently Amin (2008) has argued in favour of an "agnostic reading" of public space, urging the questioning of the assumption (widespread among urbanists) of the existence of a strong relationship between civic culture, political formation and urban public space, arguing that the latter is nowadays only a component of secondary importance in a variegated field of civic and political formation. In doing that, he does not intend to deny urban public space a role in shaping behaviour, but rather to argue that the dynamics involved are far too complex to allow a simple equivalence between "the sociology of public gathering" and "the politics of the public realm" (pp. 5–7).

The question of communication requires further clarification in this context. In fact, while communication is a defining feature of the public sphere, it is not necessarily so for public places, which are not always characterised by actual, intense verbal communication among co-present people. For many urban dwellers, communication face-to-face is indeed something that occurs more often in private than in public places. However, what sets the latter apart is the fact that *potential* for face-to-face communication *among co-present strangers* is there – whether it really occurs or not.

In accordance with this position, I will refer in this chapter to a notion of urban public space interpreting it more in terms of social response to anonymous others, than in terms of its centrality as a site of civic and political formation. I will proceed from the understanding of public space as the one which is mainly defined by the possibility of co-existence with a potential of interaction – but also with the potential of the lack of it (cfr. Pachenkov and Voronkova, this volume).

(p. 122). The present chapter on the contrary aims to show how the distinction between private and public space may still shed light on the mobility choices of urban dwellers.

2   Tomas describes how the French edition of "Strukturwandel der Öffentlichkeit" was published under the title "L'espace public" (instead of the more appropriate "La sphere publique") in 1978, at a time when the debate concerning the crisis of public space was lively (2001, p. 76).

Even more than the concept of public space, the notion of mobility has been stretched to various lengths and in different directions. At one extreme, many social scientists and geographers share a narrow understanding of the concept as "... spatial, physical mobility which only looks at brief periods of study, mostly focusing on daily travel routines and behaviour" (Nadler, 2009, p. 8). At the other end of the spectrum, there are those social theorists who have equated the contemporary phase of modernisation to mobility, using a variety of concepts that refer to movement and motion in order to point at "... the inherent thrust of modernity to 'liquefy' and 'mobilise' human and non-human agents equally, globally", setting them free from their local contexts, mainly as a consequence of the action of new transportation and communication technologies (Beckmann, 2000, pp. 17–18).

A good example of this latter position is expressed in Urry's book "Sociology Beyond Societies – Mobilities for the twenty-first century" (2000), which is defined by the author as "... a manifesto for a sociology that examines the diverse mobilities of peoples, objects, images, information and wastes" (p. 1). In this context, the term "mobility" points at things as diverse as: corporeal mobility through different modes of transport; the movement of various kinds of objects all over the globe; imaginative travel through media such as television and radio; virtual, instantaneous interactivity through ICT technologies such as the Internet (pp. 49–76). Urry's argument aims at showing how the paradigm of modern sociology, focused on the concept of distinct, self-reproducing "nation-state-societies", is today undermined by the growing importance of various global "networks and flows". In doing that, however, the term "mobility" is stretched to such a point that it can only be defined, in the most general sense, as "... an abstract concept generally describing movement in very different ways and thematic fields" (Nadler, 2009, p. 2).

In between these two extreme definitions, there are many concepts and fields of research that may or may not be subsumed by the concept of mobility, depending on the definition chosen. For example, migration and residential mobility could be seen as part of the "spatial mobility" field of research, and even social mobility (defined as movement across social positions) could be included in the all-encompassing definition cited above (Nadler, 2009).

Of course, the variety of ways in which the term is used in the social sciences is not a problem as such but, as Hannerz once stated, "... concepts are tools for thinking, and unless we know how to use them our thinking may be bad" (1969, p. 178). For example, when scholars do not specify clearly what definition they are using, this implicit conceptual mixture may lead to confusion. It may happen, for instance, that conclusions drawn from one level of analysis (for example, urban mobility) are reflected on a more general level (globalisation, modernity, etc.) or vice versa, without much if any argumentation. The consequences may be even more harmful when mobility is considered in relation to public space: in this case, in fact, the potential for vagueness is on both sides of the relation, as I have tried to show in this section. For example, when writing about the effects of mobility on public space, it is important to state very clearly whether the focus is on the consequences of globalisation on the public sphere as a site of political participation, or else on the impact of increasing urban mobility on public places such as streets, squares and the like[3].

In the present chapter, I will stick to the definition of mobility as "... short-term, repetitive and technologically and socio-organizationally constrained movement (of people) across space within daily action spaces", common in the studies at the convergence of social science and geography (Nadler, 2009, p. 6). In particular, in the next section I will try to show how the significance of the public spaces of public transport has been often neglected, in favour of the widespread view of a conflicting relationship between mobility and public space.

---

3 Almost every combination is possible, at least in theory.

Mobility and urban public space: a contentious relationship?

In general, the relationship between mobility and urban public space has often been depicted as very adversarial. According to Tomas (2001), when in the 1970s the term 'public space' established itself as a key concept in urban studies, it was partly conceived "... as a way to announce and denounce its own decline" as well as, at the same time, a way to react to it (own translation) (pp. 77–78). This degradation has happened – in the view of many scholars – "... for the benefit of movement, and more particularly of the motor car"[4] (own translation) (p. 77).

The advent of motorised transportation has indeed meant an erosion of urban public spaces, in at least four directions[5].

First, before the beginning of the 20th century, mixed use of the street – which allowed equal access to all users – was customary. Introduction of the automobile led to battles over the use of street space, which ultimately resulted in the isolation of roads from pedestrians, mainly for safety reasons (Norton, 2008). Beckmann (2000) has argued that the expansion of automobility has been endowed with many restrictions and immobilizations for other modes of transport (such as walking and cycling) because it has created prohibited places that cannot be accessed by non-drivers. In that sense, throughout the last century, the motor car has eroded urban public space through the appropriation of city streets, a large number of which have been deprived of their past functions and turned into car-only environments[6].

A second process, closely related to the first, has to do with the assumed shrinking amount of public space available in urban contexts, brought about by the increasing supply of roads and parking spaces, often provided with the goal of eliminating congestion[7]. In fact, the automobile disputes the same space with

---

4   Sennet has argued for example that "today, we experience an ease of motion unknown to any prior urban civilization ... we take unrestricted motion to be an absolute right ... the effect on public space, especially the space of the urban street, is that space becomes meaningless or even maddening unless it can be subordinated to free movement" (1977, p. 14).
5   I will propose a fifth one at the end of the chapter.
6   According to Urry "about one-quarter of the land in London and nearly one-half of that in Los Angeles is devoted to car-only environments, where in a sense the public spaces involved in urbanization have been swamped by automobility" (2000, p. 193).
7   Although now widely discredited in scientific circles, as scholars and planners have come to recognize that large increases in capacity can exacerbate congestion problems instead of solving them, this "predict and provide" approach is still periodically

other uses, and its amazing proliferation has often meant that public spaces such as sidewalks, squares, and green areas have got the worst of it.

Third, the spreading of the system of automobility (see Urry, 2004) has been tightly interwoven with the process of suburbanisation, and suburbs have often been regarded as lacking in public space, but abounding with areas devoted to the circulation of cars (Tomas, 2001, p. 80). The automobile has indeed allowed low density sprawl around cities, and residential densities at the block level tend to determine the balance between private and public space within walking distance of the home (Kemeny, 1992, p. 159). So, in a sense, the car has been held responsible also for the poor provision of public space in many new suburban areas.

Fourth, it has been observed that highways, freeways and other car-only environments often form architectural barriers that tend to fragment urban geography (Beckmann, 2000, p. 20), acting like "border vacuums" that cause urban blight in adjacent areas, often constituted by stagnant public spaces (Jacobs, 1961).

For all of these reasons, mobility as such has often been depicted as inimical to urban public space, assuming implicitly a rather static notion of the latter. Likewise, as Merriman has pointed out, "spaces or landscapes of travel and mobility are frequently referred to as being 'placeless', 'abstract', 'a-geographical', 'non-places'"[8] (2004, p. 146). In this chapter I argue that this view is limited and needs to be refined in several ways. First of all, as some scholars have pointed out, mobility is integral to the existence of public places; in fact, in most definitions of public space the 'diversity' of people and the presence of strangers is paramount and mobility, as Paulos and Goodman observe, "... is a key factor in the existence of strangers. ... The stranger, by definition from *elsewhere*, represents mobility" (2004, p. 225 – italics in the original). Similarly, Amin has argued that the peculiarity of public space is to be found in its "situated surplus", a term by which he means spaces that consist of many different things, activities and impulses that "... do not form part of an overall plan" and "... constantly change the character of the space"; thus, in his definition, the emphasis is on the "continual flux" and the "entanglements of bodies in motion" typical of public space, rather than on attributes of immobility (2008, pp. 10-11). At a more general level, Sheller and Urry (2000) have emphasised the role of mobility (and automobility in particular) in enabling

---

reiterated in political circles and among the general public (Porta, 1999, p. 446; Henderson, 2006, p. 295; Newman and Kenworthy, 1999).

8   In particular Augé's concept of "non-places" (1995) is very often used – perhaps even against the author's original intention (see Merriman, 2004) – to convey the idea that places of movement are alienating, inauthentic and devoid of meaning as such.

people to maintain networks of relationships, accumulate social capital and even reach public spaces (Albertsen and Diken, 2001), while Merriman has argued for "... a more open, inclusive working of place, in which the movement of travellers (is seen as) integral to the construction and performance of landscapes and places" (2004, p. 146).

For the purpose of this chapter however, two other, more specific remarks are paramount.

First, what has gone largely unnoticed in the common view of a contentious relationship between urban mobility and public space is the fact that other means of transport, besides the motor car, do still exist. Although automobility is certainly the predominant global form of mobility (and the number of cars worldwide is growing steadily) it has to be recognised that in many cities, including the car-dependent EU, *public transport* still accounts for a considerable share of journeys[9]. To put it another way: the adversarial view of the relation between mobility and urban public space is – at least in part – founded on a disregard for the existence of modes of transport alternative to the car, in general, and of public transport in particular.

Second, and more importantly, the fact that car users and public transport passengers spend their travel time in quite different kinds of place has largely been ignored. While the interior of the automobile may be defined as a (semi) private space, a journey by public transport usually involves the crossing of public spaces such as the bus stop, the station platform, and the subway car. In the next section, I will try to detail how the differences between the two kinds of space can be conceptualised more in depth.

Car and public transport: two different kinds of space

While the spaces of public transport have raised relatively little attention, researchers and theorists have shown interest in the analysis of the space of the private car.

The nature of a private car is in fact rather ambiguous; on one hand, it is certainly a private space, where a certain kind of domesticity is reproduced[10]:

---

9   About 60 billion passenger journeys were made by public transport in 2008 in the EU-27, representing about 120 public transport journeys per inhabitant per year (300 in the medium and large sized cities). Moreover, public transport ridership appears to have increased steadily in the last 10 years in many EU countries (UITP - International Association of Public Transport, n.d.).

10 In this article, for the sake of simplicity, I do not deal with those practices, such as carsharing, car pooling, car clubs and car-hire schemes, that tend to de-privatize the car (Urry, 2004, p. 34). Let it just be said that while a greater de-privatization of the car may

"a living room on wheels", as a Ford brochure declared in 1949[11] (Urry, 2006, p. 23). On the other hand, it must be noted that car drivers and passengers are to some extent in a public place, insofar as there always exists the possibility of being seen by others through the windows or of being involved in a car accident. For this reason, Sheller and Urry define the car as a "rolling private-in-public space" (2000, p. 746), while other scholars have tried to convey this ambiguity through the metaphor of the bubble, which provides a notion of inviolability and protection but "... in other senses ... is nothing if not fragile, liable to be popped at any moment" (Mitchell, 2005, p. 78). Furthermore, the pervasiveness of mobile communication technologies makes it increasingly possible to engage in private communication while driving, thus making the car, even if solely occupied, a potential site for meaningful interaction with distant others (see Bull, 2004; Laurier, 2004; Flamm, 2005; Urry, 2006).

Should we for these reasons conclude that the boundaries between private and public transport are blurring, and that the experience of travelling by car is somehow becoming more 'public'? Probably not. What I argue in this chapter is in fact that the automobile and public transport still are, to a large extent, on different sides of the public-private space divide. I will make this point clearer by discussing more in depth *four dimensions of travel experience* that present themselves differently in the two modes of transport (summarised in Fig. 1). In doing so, I will try to bring together literature contributions, quite disparate, that have focused either on the car or public transport or public space *per se*, but never on the three of them simultaneously.

	Private car	Public transport
Experience of diversity	Generally absent	Generally present
Expectation of privacy	Rather large	Limited
Co-presence with strangers	Generally absent	Defining character
Control over surrounding space	Full	Very limited

Fig. 1 Relevant dimensions of travel experience across the public-private divide

---

    be foreseeable in the future, as of today these practices still constitute a rather marginal phenomenon.
11   In the field of environmental psychology, Fraine and colleagues have shown how drivers' relationship with their cars can be associated with territoriality, and particularly how the way in which they speak of the car suggests that for them it corresponds mainly to a primary territory such as the home (Fraine et al., 2007).

First, the control over surrounding space, that is very limited on public transport, is conversely quite extended inside the car. As Urry has pointed out, up to the Second World War the car driver 'inhabited the road', "... being part of the environment in which the car travels", later developments of technologies of insulation and control systems facilitated a domestic mode of dwelling within the car (2006, p. 27). In particular, sole occupancy of the vehicle seems to "... permit the driver to have enhanced feelings of control and management of their environment" (Bull, 2004, p. 249).

Second, while there can be a rather large expectation of privacy inside the vehicle of the automobile, this can be only very limited on the means of public transport, just like in any public place. The concluding chapter of Watson's recent book on public space and urban encounters (2006) offers some interesting clues in this direction, by quoting some comments made in response to a survey question as to what people would do in private but not in public. Interestingly, quite a few of them show respondents spontaneously putting forward the car and public transport as paradigmatic examples of, respectively, private and public space[12]. It seems then that the assumption of larger expectations of privacy in the car environment than on public transport is quite in tune with the common experience of ordinary people.

Third, one of the distinctive features of the space of public transport is the inevitability of interaction – albeit minimum – with fellow travellers. Necessity to travel with the general public is in fact the core premise of the public transport experience, and it implies having to respect rules of social exchange such as the etiquettes of co-presence, as well as the complex task of respecting the private space of others in conditions of enforced proximity (Stradling et al., 2007, p. 291, Nash, 1975, p. 118). Co-presence with complete strangers in the vehicle of the car is by contrast largely a very residual experience in car travel, limited to taxis, hitch-hiking or car-pooling practices; as of today, it is definitely not a defining feature of what "travelling by car" means to most people in 'western' countries.

Finally, the potential diversity of the people who inhabit the spaces of public transport is yet another typical feature of public space, which is by definition open to the general public, regardless of individual characteristics (in terms of

---

12 The following quotation is particularly significant in this respect: "in the car I listen and sometimes sing along to music. I eat mints and sometimes fart. The latter I wouldn't do in a public space ... In public spaces you are seen by others, and certain standards of behaviour used to be observed, which are now disregarded and people behave worse in public than they do in private. I refer to public spitting and littering, both in the streets and on public transport. On the tubes fellow travellers appear to eat entire meals and then leave their packets ... on the seats and on public transport" (Watson, 2006, p. 162)

socio-economic level, ethnicity, gender and so on). Needless to say, this is not the case for the private car: the car owner/driver is in fact generally assumed to have the right (and the power) to establish who can get inside the vehicle; as a result, while automobiles may actually be sites of interaction, this is most likely to happen with family members, friend and acquaintances. In that sense, they certainly cannot be defined as places where people usually get in touch with diversity.

To put it briefly, then, the reason why cars should be considered as a (semi)private space, in contrast with public transport, is the fact that "sharing space with strangers", which is quite a common experience in the latter, is largely absent in the former[13].

This leaves out the question of communication. At the first sight, the fact that actual communication is more likely to take place inside the car than on public transport may appear paradoxical. At the same time, it has to be recognised that a face-to-face communication *with strangers* is certainly more likely to occur on public transport. This communication is often of the weakest nature: in fact, passive spectatorship and the effort to preserve the borders of personal privacy in public are often the prevalent features of the public transport experience. Even "*flâneur*" attitudes (see Nuvolati, 2006) and the propensity to the observation of the diversity of strangers gathered in public transport spaces – to be discussed at more length in the next section – may often seem to boil down more to voyeurism and passive spectatorship than a result in full-blown forms of verbal communication. Nevertheless, I agree with Newman and Kenworthy (1999, p.46) when they point out the crucial role of "unplanned access", i.e. "accidental" casual interactions (that may occasionally evolve into something more significant), in defining public transport and at the same time probably strengthening the sense of belonging to a certain place and/or social unit. It is this feature, i.e. the possibility of co-existence with strangers with a *potential* of face-to face interaction (but also with the potential of the lack of it) that, according to the definition of public space given above, sets the latter apart from private space. Of course, there can be potential without significant interaction actually taking place, but definitely not the other way round.

In that sense, public transport may indeed be considered as a kind of public place, although the quantity of communication there is relatively lower (and the quality of it, quite different) than in the car.

---

13   Of course car drivers and passengers share the space of the road with other vehicles, and consequently with other people. However, this is obviously not the same, insofar as the car provides at least a 'bubble' of private space in the middle of the public road.

Sharing space with strangers or not?
Two different kinds of attitude

As argued in the previous section, the choice between using the car instead of public transport (or vice versa) is a choice between spending travel time in two quite different kinds of space, with some contrasting features. These latter seem in turn to be judged in different, and sometimes opposite, ways by different people. Several scholars have noticed one or the other of those (cited below), and many have also analysed them in depth. However, at least as far as this author knows, none has yet tried to bring these disparate insights together in a single, coherent theoretical model. That is precisely the aim of the work presented in this chapter: the present section starts then by reviewing the various contributions and trying to put them in perspective in a unitary way, while the next section will present the (tentative) theoretical model, together with some empirical results.

Various studies have shown how many people highly value the privacy of the car, which makes it possible to engage in private communication with distant

others through mobile devices (Urry, 2006, p. 28), to listen and sing along to music (Bull, 2004), and even in some instances to "do office work on the motorway" (Laurier, 2004). In that sense, the private nature of the interior of the automobile seems to allow a certain variety of modes of appropriation of travel time (Flamm, 2005) and, according to Bull, the sole occupancy of the car "... permits the driver to have enhanced feelings of control and management of their environment" as well as providing "... spaces of temporary respite from demands of the 'other'" (2004, p. 249). Other authors have stressed by contrast how the automobile can facilitate a domestic mode of dwelling also by producing an environment in which a certain sociability with present others can occur (Urry, 2006, p. 27): cars may in fact represent an important site for family life, intimate interactions (Sheller, 2004) and particularly for "cultures of mothering" (Dowling, 2000). In that sense, the car may be considered as a space where the social reproduction of groups (not necessarily just families) and social identities takes place through communication. Arguably, the degree of isolation from the outside world and the general public plays a crucial role in fostering sociability inside the car, especially if it is of intimate nature.

On the contrary, as argued in the previous section, to travel by public transport necessarily means having to share space with strangers and to put on a "commuter stance", which "... is highly institutionalised and involves the negotiation of private space or territory within the private domain" (Nash, 1975, p. 119). This can be experienced as an annoyance: environmental psychologists and proxemics scholars have shown in fact how physiological stress elevates in conditions of crowding, when intrusions into personal space multiply (Hall, 1966, p. 118; Evans and Wener, 2007). Rocci has documented through qualitative interviews how the co-presence with strangers on the Parisian subway is consciously experienced by many passengers as a source of stress, aggressiveness and promiscuity (Rocci, 2007, p. 70).

However, and this is a key point, there is another side of the coin: in fact, there are good reasons to believe that many consider a ride on the bus as an opportunity to observe or participate in social interactions with fellow passengers, and ultimately see it as one of the psychological benefits of public transport. This point of view has been suggested, for instance, by recent research work by Stradling and colleagues on passenger perceptions of the urban bus journey experience in Edinburgh (2007, pp. 289–291). Of course, as already mentioned in the previous section, "strong" forms of interaction such as starting a conversation with a stranger are probably relatively rare, and more likely to take place in particular contexts such as long distance buses and trains[14]. By contrast, "soft" forms of interaction such as observing other passengers or listening to their chatter are probably more likely to occur, as pointed out by Paulos and Goodman (2004). They have urged scholars to take into account our relationship with "familiar strangers", i.e. those individuals that we repeatedly observe in public urban spaces and yet not directly interact with, whose best example is "... a person that one sees on the subway every morning" (p. 223)[15].

Italian urban sociologist Giampaolo Nuvolati's work on urban *flâneurs* (2006) may be a useful reference in this context. Nuvolati's *flâneur* is a peculiar type of urban dweller who loves to mix with the crowd, especially in places characterised by a state of constant flow, such as sites of consumption, mobility and waiting. He (or she) is particularly attracted by the interpenetration of house and street, public and private that defines the crowd, since that provides the possibility to observe without being observed, to mix with strangers without losing anonymity and to perform the creative, interpretative acts that are often required to process the *stimuli* coming from crowded public spaces (that are, by definition, unpredictable and problematic). However, Nuvolati warns that:

> direct relationships with the resident population are rather occasional, because the *flâneur* ... doesn't have any need to develop strong bonds with the actors on the scene. Most of the time the relationships are either instrumental or aimed to accelerate the decoding of the *genius loci*. (own translation) (p. 118)

In that sense, the attitudes exemplified by the *flâneur* are not without relation with contemporary forms of extreme subjectivism and narcissism, even if Nuvolati concedes that in the case of sedentary *flâneurs* (those who observe and explore their own city), a higher level of interaction and exchange with the local reality and a more relevant emotive component should be expected.

Nuvolati's remarks are useful in this context because they seem to suggest that *flâneur*-like attitudes may be relevant in the daily lives of many urban

---

14 Nash, in his ethnography of bus riding, has pointed out three structural conditions that encourage the transformation of buses in "arenas for social encounters": the degree of competency of riders, a low number of people in the vehicle, and a long duration of the ride (1975, p. 119).

15 The notion of "familiar stranger" was actually coined by psychologist Stanley Milgram in 1972, who investigated the phenomenon through an experiment whereby his students photographed people waiting on the platform of a suburban light rail station, returned at the same time of day a week later, and asked people to label individuals they recognized in prints of the photograph: an astonishing 89% of the sample recognized at least one person (Milgram, 1977, cited in Paulos and Goodman, 2004, p. 223). Paulos and Goodman have simply replicated the experiment in Berkeley, finding lower (77,8%) but still high recognition (2004, pp. 226–227).

dwellers: thus, it may be sensible to assume that they are common also among public transport passengers. In the latter case the propensity to observe other passengers and to listen to their chatter may be intense, but this doesn't necessarily entail a motivation to speak or become friends with them.

While this is important, it should not lead us to obscure the fact that public transport has indeed the potential for interactions, ranging from the mere observation and recognition of co-present strangers to full blown social exchanges. In that sense, it seems to share a quintessential quality with more 'traditional' public spaces such as city streets (Stradling et al., 2007, p. 291).

Another peculiar trait of Nuvolati's *flâneur* may be of interest here: he is particularly attracted to the diversity of people that may be encountered among the crowd, as he aims to "... capture and tell the most daring encounters between different populations" (own translation) (p.43), notably if they involve marginalised social groups and individuals that question the established order. In that sense, it may be hypothesised that some passengers, notably those inclined to enjoy the observation of the crowd, may actually like the potential diversity of the people on public transport, and see also that as one of the psychological benefits of this mode of transport.

Of course, this is unlikely to hold true for everyone, as others would probably tend to experience diversity as an annoyance. This attitude is particularly apparent in the following quote of John Rocker, a young white male baseball player from Atlanta (one of the most sprawling and car dependent cities in the United States), whose comments about New York's subway published on "Sports Illustrated" sparked much controversy in 1999 (Henderson, 2006, p. 297):

> imagine having to take the number seven train to the ballpark, looking like you're riding through Beirut next to some kid with purple hair next to some queer with AIDS next to some dude who just got out of jail for the fourth time next to some 20-year old mom with four kids. It's depressing. (Pearlman, 1999)

Public transport has in particular often been identified – as the quotation above shows with particular clarity – with marginal, stigmatised or minority groups. This is more likely to happen in especially car dependent urban areas, where public transport is used mainly by those who cannot afford a car or do not have a driving license. In that context, the choice to use the car instead of public transport might be reinforced by the will to avoid people belonging to such groups. Henderson has tried to shed light on this process by putting forward the notion of "secessionist automobility", or "... using the car as a means of physically separating oneself from spatial configurations like higher urban density, public space, or from the city altogether" – while at the same time seeking to "... avoid people of other races or classes, or to avoid spontaneous interaction on public streets"[16] (2006, pp. 294, 296). In that sense, the notion of automobility as a tool of spatial secession proposed by Henderson aims to show how the car provides for whites in Atlanta "... a means of travel through spaces inhabited by blacks, all without having to interact with blacks" (2006, p. 299). Thereby it can be considered as a kind of "mobile" counterpart to the far more established concept of (residential) urban segregation, insofar as it posits that social mixing and social avoidance occur not only in neighbourhoods, but also in the flows of urban and suburban mobility[17].

Watson (2006), has put forward an interesting explanation for the distaste that some feel towards the diversity of people in public places, that may be of particular interest for the study of attitudes towards public transport. Her argument is that:

> ... underlying the resistances to rubbing along in public encountering others who are different, is a distaste towards others who behave in ways that are deemed inappropriate or unacceptable, often because they are designated as 'private', and this produces and legitimates hostility in the self towards others who are different. (p. 161)

The point is in fact that "different cultures have different understandings of space and the kinds of embodied practices which are appropriate or not in public" and "... this can operate across racial/ethnic differences as well as those of age ..., or across gender and sexual orientation" (pp. 165–166). I argue here that as a result of the processes described by Watson, the diversity of people on public transport may be perceived as an annoyance by some travellers, for reasons a little more complex than mere xenophobia.

It should be apparent, after the brief literature review presented in this section, that there are many good reasons to assume that travellers diverge in their degree of enjoyment of the typical features of public space (such as the co-

---

16 While remarkable, the concept of "secessionist automobility" put forward by Henderson is mainly intended to shed light on the politics of automobility – defined as "... a spatial struggle over how the city should be organized and for whom" (2006, p. 293) - rather than on the micro-level individual choices, behaviours and attitudes of urban dwellers. In that sense, it is quite different from the dimension social mixing-secessionism that will be proposed in the next section, even if I chose to use the same expression in order to avoid contributing to a useless proliferation of terms.

17 Of course, the quotation of John Rocker, as well as the whole concept of "secessionist automobility" put forward by Henderson, should be judged against the particular level of residential ethnic segregation in U.S. cities in general, and in Atlanta in particular (see Thompson, 2000). However, I argue that both can be a useful reference insofar as they show in extreme (and therefore clearer) forms processes that may be at work also in different and less segregated contexts.

presence and proximity with strangers, and the diversity of people) that characterise the spaces of public transport. Similarly, the main features of the car environment (isolation, possibility of sole occupancy, absence of diversity, etc.) may be judged in contrasting ways by different people. These diverging attitudes towards these two extreme modes of transport deal rather clearly with the values of privacy and publicness in contemporary urban life, and should thus be of interest to scholars concerned with the relations between public space and mobility in cities. At the same time, they should draw also the attention of transport scholars, since it may be assumed that these diverging views may even contribute to leading people to different "modal choices" of the transport means[18].

I propose then to link explicitly the liking for the private space of the car to the disliking of the spaces of public transport, by putting forward a tentative single attitude dimension opposing *social mixing* and *secessionism* in mobility that intends to deal with both sides of the question in conjunction. This notion, to be analysed in detail in the next section, is intended as a useful conceptual tool to tackle in a unitary way the various questions analysed by the disparate literature reviewed above. At the same time, it opens up the opportunity to explore these issues more in depth by conducting an empirical enquiry.

Social mixing and secessionism in mobility: empirical research of attitudes

The academic literature on urban mobility, notably in the field of social psychology, has often assumed that social attitudes are worth considering in order to understand mobility behaviour and, more specifically, modal choice[19] (see for example Garvill et al., 2003; Fuji and Kitamura, 2003; Bamberg et al. 2003). However, these studies have generally tried to measure the degree of liking a certain mode of transport (mostly the car) *per se*, rather than focusing on

---

18 The term "modal choice" (or "mode choice") is common in the field of transport engineering, where it is used in transportation forecasting models (see Edwards, 1992). Here I use the concept in a rather minimal and non specific way, meaning the choice an individual makes between different modes of transport for a specific journey. Of course, "choice" itself is sometimes a misleading term: in many cases in fact there is no actual choice, and it is more a matter of habitualized practice, as many studies in the field of social psychology have clearly shown (see, for example, Gärling and Axhausen, 2003).

19 Of course, factors other than social attitudes are definitely more relevant in influencing modal choice. Among them: urban structure, availability of transport infrastructure, place of residence, personal capabilities, income, car ownership, etc.

more specific attitude dimensions that may have an impact on modal choice[20]. Even when this latter effort has been made (see for example Golob and Hensher, 1998; Anable, 2005), scholars have generally omitted the significance of the varying propensity of individuals to share space with strangers during travel. By contrast, in this section I will explore the utility of conceiving this continuum as a hypothetical attitude dimension opposing *social mixing* and *secessionism in mobility*, thus linking on a single dimension the contrasting ways in which travellers value the spaces of the two modes of transport described in the previous sections[21].

On the basis of what has been argued in the previous section, I broke down this attitude continuum into three main sub-dimensions (see Fig. 2): first, the co-presence and interaction with strangers on public transport, which should be disliked by secessionist subjects (who should prefer to avoid it through the use of the car) but enjoyed by subjects at the other end of the spectrum, who should like activities such as being in public and observing the crowd. Second, pro-mixing travellers should enjoy the diversity of people on public transport, whereas secessionist ones should dislike it, especially if marginal or minority groups are represented. Finally, sole occupancy of the car should be a pleasure for subjects at one extreme of the continuum, particularly enhanced by the use of sound and communication devices, whereas pro-mixing subjects should dislike it and feel it as a form of seclusion.

---

20 As a matter of fact, the relationship between attitudes and behaviour is always very ambiguous and difficult to disentangle at best. The impact of behaviour on attitudes may in fact often be at least as relevant as the opposite, as a result of processes such as cognitive dissonance reduction (Festinger, 1957). For instance, as far as the attitude dimension proposed here is concerned, it may be argued that owning a car or living in the suburbs brings about car use, which in turn may lead to the development of attitudes that have been labelled in this chapter as "secessionist". However, it may also be posited that attitudes, once established, influence in turn behaviour, by "freezing" it and making it resistant to change. Golob and Hensher (1998) have proven by means of structural equation modelling that attitudes about the status-symbol effect of the car and congestion are linked in a feedback loop with the choice of solo-driving and (un)willingness to reduce vehicle kms. In this perspective, obtaining a measure of the attitude dimension secessionism-social mixing could also be useful to test such conjectures about the relationship between modal choice and attitudes.

21 The attentive reader might be surprised by the lack of consideration given to other modes of transport, such as bicycles and motorbikes. The choice to focus exclusively on the car and public transport can be justified, on the one hand, by their prominent importance in many systems of urban mobility and on the other hand, by the fact that they are perfect opposites on the private-public continuum, as argued in the previous section. In that sense, other means of transport can be imagined as placed in between these two extremes, as it will be argued in the conclusion of this chapter.

	Secessionism	Social Mixing
Co-presence and interaction with strangers on PT	Disliked, propensity to avoid it through the use of the car	Liking of activities such as being in public, observing the crowd, etc.
Diversity of people on PT	Disliked, particularly in relation to marginal or minority groups	Enjoyment of diversity
Sole occupancy of the car	Liking, enhanced by the use of sound and communication devices	Disliked, perceived as a form of seclusion

Fig. 2: *The main sub-dimensions of the secessionism-social mixing continuum*

Of course, as long as no empirical evidence is delivered, this attitude dimension would simply remain a working hypothesis (perhaps over-simplified), based on the gathering of scattered observations by authors in very different disciplinary fields, (almost) none of which ever argued that such a dimension could exist. In order to test this hypothesis then, a 28-items Likert scale has been generated, with the sub-dimensional structure presented in Fig. 2 in mind (a few examples of the items are presented in Fig. 3), and pre-tested on a small sample (n=12) stratified by age, gender and education level (all respondents had a driving licence)[22]: a little-used pre-testing procedure, called respondent debriefing, was carried out in the Milan area in 2009. The main goal of the procedure was to ensure that the meaning of the item as written in the questionnaire was consistent with the way respondents interpret it (Hughes, 2004). This has meant conducting standardised interviews with follow-up (open) questions (in-depth

---

22 Scales are often presented as a final product, while the process of scale development is often overlooked, and only rarely discussed. I think on the contrary that it is worth some attention, and thus I proceed to describe it here briefly; to create the scale, I have in fact undertaken the following steps, as suggested by the literature on attitude scaling (see DeVellis, 2003): in the first place, I have generated a large item pool (41 items), with a certain degree of redundancy. Sources for the items have been: previous qualitative studies, press articles and the like and existing attitude scales. This initial item pool has been reviewed by four experts (in the fields of mobility, sociology, attitude scaling and linguistic), leading to a first selection, after which only 28 items have been retained.

	Secessionism	Social Mixing
Co-presence and interaction with strangers on PT	"A thing I don't like about public transport is being forced to share space with strangers"	"It is interesting to observe people on public transport"
Diversity of people on PT	"There's too many weird people on public transport"	"One can always see interesting people on public transport"
Sole occupancy of the car	"It is enjoyable to drive alone"	"A thing I don't like about driving alone is that I feel isolated from the outside world"

Fig. 3: Examples of items in the scale "secessionism – social mixing in mobility"

probes): in substance, respondents were first asked to fill in the questionnaire, composed of 5-point Likert-scale items and then, for each item, to explain "why" they had chosen to rate it as they did.

The aim of the respondent debriefing was mainly methodological: with respect to this purpose, the procedure has proved useful, leading to a second selection (six items were dropped), and to changes in wording for many of them. Nonetheless, and even if it is a very preliminary stage of the planned empirical research, the material gathered through the interviews can be used here to illustrate the significance and flaws of the proposed attitude dimension, and to formulate some tentative hypothesis about the articulation of its structure, which is likely to be more complex than first thought[23].

From the outset, it should be recognised that a few respondents have indeed put forward the ambiguity of the space of the car; in doing that, they have referred mainly to the transparency of windows (which entails the possibility to be seen by others outside the car and, conversely, to "have a look around") as well as to the presence of road traffic[24]:

---

23 A proper assessment of the internal consistency of the scale and of its structure requires of course the conducting of a pilot testing on a large and representative sample, allowing the subsequent carrying out of an item analysis, and the finalization of the scale through a further selection of the items.
24 However, the role of road traffic appears to be somewhat ambiguous in itself: other respondents have pointed out in fact that it requires constant attention, thus further isolating the driver from the 'outside world' and contributing to the privateness of the car environment. It should also be noted that no respondent mentioned the possibility to

> ... you're in a car on the road but if in every car there's a person, it's just like being in the crowd. Of course you don't collide with them, you don't perceive the smell.. but I don't see that as a separation.. I mean, when you're in traffic it's like.. well, maybe not like being in the subway but.. more or less.. it's up to you: if you want to preclude yourself, you can stare the wheel, otherwise..(SA, 25, M, high level of education)

However, many more comments emphasise the private nature of the space of the car, sometimes comparing it explicitly to the domestic realm and using arguments that are rather consistent with the theoretical description of the attitude dimension put forward in this chapter. As a matter of fact, those respondents who have judged positively the privacy provided by the automobile have cited very often the possibility to do things judged inappropriate in public (such as eating, smoking, listening and singing along to music, making phone calls, talking aloud, etc.) and defined the time spent travelling in a car as a valuable time of pause and of 'chilling out'. Others have on the contrary defined that experience in negative terms, putting forward the lack of communication and interaction with the outside world, a sense of solitude and isolation, and the general feeling that "time does not pass as quickly" as one wishes:

> when I use the car ... I put on what I want.. I never drive alone so... we let our hair down ... in my family... we speak our dialect, with no one hearing... we say stupid things, we buzz about... a little bit like when you're at home and you do... that kind of things, exactly. (RG, 56, F, low level of education)
>
> ... I feel isolated, yes... when I close the door and get inside... I feel lonely... sometimes I do feel like being alone, no mess around me... and some other times maybe, travelling alone ... it's a burden: it seems longer, you know what I mean? (MC, 63, F, low level of education)

As far as public transport is concerned, some respondents have indeed voiced their enjoyment of the diversity of the crowd, with arguments that seem to confirm its relevance in characterising public transport as a kind of public space. The most cited observed factors of diversity are origin (national, but also regional), accent, language, dress style, social status, and hairstyle; particularly common has been the reference to foreigners and immigrants[25]:

---

engage in communication through mobile devices as a proof of the public nature of the car; on the contrary, a few of them argued that it's precisely the privacy of the automobile that makes them feel at ease in making phone calls. This last result is consistent with the qualitative study of Flamm, conducted on a sample of Swiss travelers (2005, p. 16).

25 This result should be judged against the fact that Milan has only relatively recently experienced important waves of immigration from abroad, if compared to big cities in

The automobile has indeed allowed low density sprawl around cities and in residential areas ... the car has been held responsible also for the poor provision of public space in many new suburban areas.

... interesting people, I mean... people you're not used to ... different from your circle of friends... above all, you realise what there is in your town. I mean, if you only travel by car, probably you have an altered perception, I mean... Not everyone has a car, if you take the public transport you really see everything... everything that goes around in town. (SA, 25, M, high level of education)

... you see the weirdest people... the most incredible faces, I mean... I never thought to see that in Bangkok or Manila, and now I find it in Milan... that makes me curious, just like being abroad... hearing those people speak weird languages, it's a thing that fascinates me[26]. (GP, 67, M, low level of education)

Accordingly, several comments highlight also the fact that activities such as observing the crowd on public transport are quite common among respondents, even if some others tend to judge this opportunity with scepticism, putting forward a rather 'impersonal' view of travel time on public transport:

---

other western European countries; however, immigration from southern Italy has been massive during the second half of the 20th century (see Foot, 2001).

26   It is worth notice that no respondent has voiced dislike for the diversity of people on public transport or towards the presence of marginal or minority groups; many clues seem to suggest however that social desirability response bias has been at work in producing this result, probably exacerbated by the face-to-face nature of the interviews.

> ... I really liked watching people.. generally speaking... if you see them on public transport ... you watch their conduct, you know? You can see if somebody is doing something, if he's busy and then... I liked to do this kind of observation when I was a kid... and I still do it when I can. (GF, 60, M, low level of education)
>
> ... what we have here is a romantic notion of public transport.. I mean, what does it mean "interesting"? I may meet interesting people at a concert... when I go on public transport I don't expect to see interesting people... how could I figure it out? By looking at their coat? I have no idea ... I don't have any particular interest in relationships on... public transport. (NDL, 53, F, high level of education)

In sum, the results of the interviews conducted seem to support the idea that travelers see the car and public transport as two rather contrasting kinds of space, differentiating them along a private-public line. Besides, many comments seem to show the significance of some of the attitudes illustrated in Fig. 2, such as the enjoyment for the diversity of people on public transport and for the sole occupancy of the car. The respondent debriefing also suggests that the dimension "secessionism-social mixing" may have been somewhat too rigidly conceived: many respondents seem in fact to value both the possibilities of secession provided by the car and the opportunities for social-mixing provided by public transport; thus, I find that it is perhaps more realistic to assume the two dimensions are to a certain extent autonomous and unrelated to each other[27].

This new hypothesis (that implies the bi-dimensionality of the construct, initially conceived as a single continuum) is nonetheless interesting. In fact, the existence of a cluster of travelers who value both possibilities could be seen as a sign that there is no inherent preference towards the "privacy" of the car among contemporary urban dwellers, and that these travelers might eventually be steered towards more sustainable, multimodal or public mobility patterns, given the right conditions. In that sense, the co-presence and the diversity of people on public transport – the fact of it being a public place – could be conceived as a potential factor of competitiveness, rather than just a flaw.

Conclusion

As stated in the introduction of this chapter, much of the debate about urban public space has focused on its assumed erosion in contemporary cities. In this perspective, the discussion conducted above leads us to recognise that the

---

[27] Of course, any final word on the internal structure of the attitude dimension requires the conducting of a pilot testing of the scale, as stated above. These results are presented in Mattioli 2012 (forthcoming).

predominance of the motor car in contemporary 'western' mobility systems is detrimental to public space in a fifth, rarely discussed way: quite simply, it leads more people to spend more time in the private space of the car, rather than in the public spaces of public transport. Conversely, cities with high rates of public transport ridership should be considered, also for this reason, as favourable contexts for the endurance of public space[28]. This basic fact is even more relevant if one considers the amount of daily time dedicated to mobility, which seems to be increasing in some countries[29]. This leads to a question that should be of concern to scholars interested in the fate of public space in contemporary cities: is increasing car-dependence making urban dwellers less accustomed to sharing space with strangers and to experiencing diversity – in public transport and in public space more in general? The approach drafted in this chapter may be useful in trying to empirically answer this research question.

However, this remains a quite traditional approach to urban public space, and one that could be called into question. Some might argue, for instance, that it is necessary to go beyond the general division of private/public space and to analyse more in depth the practices, attitudes and experiences that are associated with private and public space in the daily lives of urban dwellers. The work presented in this chapter has tried precisely to do that, by providing a general framework to explore more in detail the attitudes of car and public transport users towards these two modes of transport.

In that sense, the above discussion shows clearly that it would certainly be an exaggeration to equate public transport with every conceivable virtue of publicness and, conversely, cars with antisociability and isolation. It seems in fact that cars can often be places of intense sociability, while interaction among strangers on public transport is often very weak. My point of view (and my value judgement) in this respect is that the potential for face-to-face interaction

---

28 Surprisingly, public transport companies seem little aware of these facts. For example, the position paper of the International Association of Public Transport on "Assessing the benefits of public transport" (UITP - International Association of Public Transport, 2009) considers the relationship between public transport and public space only in terms of what the former can do to reduce the volume of car traffic, thus "... allowing valuable public space to be used for walking, cycling, relaxing and enjoying our cities". A similar view is expressed in the Project for Public Spaces' "Transportation Program" (Project for Public Spaces).

29 In Italy, for example, the value of this indicator for the adult population in 2002-2003 was 1 hour and 22 minutes, which is 24 minutes more than in 1988-1989 (Colleoni, 2008, p. 101; ISTAT, 2007). At a more general level, according to Castells, the diversification of working sites, induced by new ICT technologies for a large fraction of the population, will result in an appreciable increase of commuting time, particularly in European cities (Castells, 2000, p. 426).

with strangers that sets public transport apart from automobiles (what Newman and Kenworthy name "unplanned access", 1999, p. 46), is an important and valuable feature to be preserved in contemporary cities. However, the question is of course open to further discussion and contributions, which will hopefully fuel a debate about the relations between urban mobility and public space, one that is highly needed.

In this context, more attention should probably be paid to the modes of transport located in between the two extremes on the private-public continuum: for instance, motorbikes share many characteristics with cars, but the lack of a 'bubble' makes them somewhat less private. Bicycles, on the other hand, are arguably even more public, because their slowness likely increases the degree of interaction with the surrounding environment. However, both of these modes of transport are to a certain extent private (because no 'sharing space with strangers' is required) and that possibly makes them favourite choices for secessionists who do not want to (or can not) use a car. Future research should thus notably focus on the role of bicycling, which is increasingly popular in 'western' cities struggling for sustainability, while it is probably losing primacy as a mode of transport in some former developing countries such as China.

Furthermore, the fact that public transport as a kind of public space has been given little (if any) consideration shows clearly that there is an urgent need to recognise that:

> ... public spaces take many different forms, and the more visible and documented spaces of the mall or town square, the park or the piazza, though important, are only one part of the story. (Watson, 2006, p. 170)

Interestingly, almost all of these more traditional and known forms of public space date back to a pre-industrial, mainly pedestrian kind of city. It may be argued that the features of what is considered urban public space change over time, and that the characters of mobility are a very relevant factor in defining them.

Bertolini and Dijst (2003) have put forward a useful tool to frame this question with the concept of "mobility environments", that is "... places and moments where mobility flows interconnect", where "... 'urban diversity' and frequency of human physical contact conserve a crucial role" (p. 31) and thus where the quintessential potential for interaction characteristic for urban public space is visibly present. As they write, "throughout history the most dominant transport mode was walking. In the past famous mobility environments developed on sites at which walking routes met" (p. 35). In that sense, it may be argued that the evolutions of mobility since the 19th century (that have led from the "Walking City" to the "Transit City" to the "Auto City", see Newman and

Kenworthy, 1999) have brought about new kinds of public space that deserve to be recognised as such. Bertolini and Dijst have taken a step in this direction, by pointing out the need to focus on transport nodes such as railway stations and the like. However, in their paper they largely neglect environments such as buses, subway cars and trains. On the basis of the arguments put forward in this chapter, I argue by contrast that an approach interested in mobility environments as a kind of public space should take into consideration the spaces of public transport in the broadest sense, thus including "moving" public places such as those cited above.

In that sense – and this is the final point of this conclusion – the spaces of public transport could be regarded as a possible point of convergence and cooperation for studies on urban mobility and on public space that often have remained worlds apart: this would mean moving towards that kind of interdisciplinary which is a prerequisite to understand and resolve the problems of contemporary urban public space (see Pachenkov and Voronkova, this volume).

References

ALBERTSEN, N. and DIKEN, B. (2001) Mobility, justification and the city. *Nordic Journal of Architectural Research*,14 (1), pp. 13–24.
AMIN, A. (2008) Collective culture and urban public space. *City*, 12 (1), pp. 5–24.
ANABLE, J. (2005) 'Complacent Car Addicts' or 'Aspiring Environmentalists'? Identifying travel behaviour segments using attitude theory. *Transport Policy*, 12, pp. 65–78.
AUGÉ, M. (1995) Non-places: introduction to an anthropology of supermodernity. London, New York: Verso.
BAMBERG, S., RÖLLE, D. and WEBER, C. (2003) Does habitual car use not lead to more resistance of travel mode? *Transportation*, 30, pp. 97–108.
BECKMANN, J. (2000) Heavy Traffic – paradoxes of a modernity mobility nexus. *Mobility and Transport - An Anthology*. Copenhagen: The Danish Transport Council, pp. 17-26.
BERTOLINI, L. and DIJST, M. (2003) Mobility environments and network cities. *Journal of urban design*, 8 (1), pp. 27–43.
BULL, M. (2004) Automobility and the power of sound. *Theory, Culture & Society,*21 (4/5), pp. 243–259.

CASTELLS, M. (2000) The rise of the network society: The information age: economy, society and culture, Volume I. 2nd ed. Oxford: Blackwell.
COLLEONI, M. (2008) La dimensione sociale della mobilità in Italia e a Milano. In: Colleoni, M. (ed.) *La ricerca sociale sulla mobilità urbana. Metodo e risultati di indagine.* Milan: Raffaello Cortina Editore, pp. 99–114.
DEVELLIS, R. F. (2003) *Scale development. Theory and applications.* 2nd ed. Thousand Oaks, California: Sage Publications.
DOWLING, R. (2000) Cultures of mothering and car use in suburban Sydney: a preliminary investigation. *Geoforum*, 31, pp. 345–353.
EDWARDS, J. D. (1992) *Transportation planning handbook.* New Jersey: Prentice-Hall.
EVANS, G. W. and WENER, R. E. (2007) Crowding and personal space invasion on the train: Please don't make me sit in the middle. *Journal of Environmental Psychology*, 27, pp. 90–94.
FESTINGER, L. (1957) *A theory of cognitive dissonance.* Stanford, California: Stanford University Press.
FLAMM, M. (2005). A qualitative perspective on travel time experience.*5th Swiss Transport Research Conference, Ascona, March 9–11 2005.*
FOOT, J. (2001) Milan since the miracle: city, culture and identity. Oxford: Berg.
FRAINE, G. et al. (2007) At home on the road? Can drivers' relationship with their cars be associated with territoriality? *Journal of Environmental Psychology*, 27, pp. 204–214.
FUJI, S. and KITAMURA, R. (2003) What does a one-month free bus ticket do to abitual drivers? *Transportation*, 30, pp. 81–95.
GÄRLING, T. and AXHAUSEN, K. W. (2003) Introduction: habitual travel choice. *Transportation*, 30, pp. 1-11.
GARVILL, J., MARELL, A. and NORDLUND, A. (2003) Effects of increased awareness on choice of travel mode. *Transportation*, 30, pp. 62–79.
GOLOB, T. and HENSHER, D. (1998) Greenhouse gas emissions and australian commuters' attitudes and behaviour concerning abatement policies and personal involvement. *Transportation Research Part D*, 3 (1), pp. 1–18.
HABERMAS, J. (1989) The structural transformation of the public sphere: an inquiry into a category of bourgeois society. Cambridge, Massachusetts: The MIT Press.
HALL, E. T. (1966) *The hidden dimension.* New York: Anchor Books.
HANNERZ, U. (1969) *Soulside. Inquiries into ghetto culture and community.* Chicago: The University of Chicago Press.

HENDERSON, J. (2006) Secessionist automobility: racism, anti-urbanism, and the politics of automobility in Atlanta, Georgia. *International Journal of Urban and Regional Research*, 30 (2), pp. 293–307.

HUGHES, K. A. (2004) Comparing Pretesting Methods: Cognitive Interviews, Respondent Debriefing, and Behavior Coding. *Survey Methodology*, 2004–02.

ISTAT. (2007) L'uso del tempo. Indagine multiscopo sulle famiglie, anni 2002–2003. Roma: Istat.

JACOBS, J. (1961) The death and life of great American cities. New York: Random House.

KEMENY, J. (1992) *Housing and social theory*. New York: Routledge.

LAURIER, E. (2004) Doing office work on the motorway. *Theory, Culture & Society*, 21 (4/5), pp. 261–277.

MATTIOLI, G. (2011) Scelta modale, atteggiamenti e condivisione dello spazio nella mobilitá quotidiana. *Sociologia Urbana e Rurale*, 94, pp. 103-118.

MERRIMAN, P. (2004) Driving places. Marc Augé, Non-places, and the geographies of England's M1 motorway. *Theory, Culture & Society*, 21 (4/5), pp. 145–167.

MITCHELL, D. (2005) The S.U.V. model of citizenship: floating bubbles, buffer zones, and the rise of the "purely atomic" individual. *Political Geography*, 24, pp. 77–100.

NADLER, R. (2009) *Multilocality: an emerging concept between the terms of mobility and migration*. Dipartimento di Sociologia e RicercaSociale, University of Milano-Bicocca. Available from: http://www.sociologiadip.unimib.it/dipartimento/ricerca/pdfDownload.php?idPaper=477.

NEWMAN, K. and KENWORTHY, J. (1999) *Sustainability and Cities. Overcoming Automobile Dependence*. Washington, D.C.; Covelo, California: Island Press.

NASH, J. (1975) Bus Riding: Community on Wheels. *Urban Life*, 4 (1), pp. 99–124.

NORTON, P. D. (2008) Fighting traffic: the dawn of the motor age in the American city. Cambridge, MA: The MIT Press.

NUVOLATI, G. (2006) Lo sguardovagabondo.Il flâneur e la città da Baudelaire ai postmoderni. Bologna: IlMulino.

PAULOS, E. and GOODMAN, E. (2004) The familiar stranger: anxiety, comfort, and play in public places. *CHI 2004, Vienna, April 24–29*, pp. 223–230.

PEARLMAN, J. (1999). John Rocker. *Sports Illustrated*, 21th Dec.

PORTA, S. (1999) The community and public spaces: ecological thinking, mobility and social life in the open spaces of the city of the future. *Futures*, 31, pp. 437–456.

PROJECT FOR PUBLIC SPACES (n.d.) *About the PPS transportation program* [WWW]. Available from: http://www.pps.org/transportation/approach/ [Accessed 06/04/2010].

ROCCI, A. (2007) De l'automobilité à la multimodalité? Analyse sociologique des freins et leviers au changement de comportements vers une réduction de l'usage de la voiture. Le cas de la région parisienne et perspective internationale.UnpublishedThesis (PhD). University Paris 5 – René Descartes – Sorbonne.

SENNETT, R. (1977) *The fall of public man*. London and Boston: Faber and Faber.

SHELLER, M. (2004) Automotive emotions: feeling the car. *Theory, Culture & Society*, 21 (4/5), pp. 221–242.

SHELLER, M. and URRY, J. (2000) The city and the car. *International Journal of Urban and Regional Research*, 24 (4), pp. 737–757.

SHELLER, M. and URRY, J. (2003) Mobile transformations of 'public' and 'private' life. *Theory, Culture & Society*, 20 (3), pp. 107–125.

STRADLING, S. et al. (2007) Passenger perceptions and the ideal bus journey experience. *Transport Policy*, 14, pp. 283–292.

THOMPSON, M. A. (2000) Black-white residential segregation in Atlanta. In D. L. Sjoquist (Ed.), *The Atlanta Paradox*. New York: Russell Sage Foundation.

TOMAS, F. (2001) L'espace public, un concept moribond ou en expansion? *Géocarrefour*, 76 (1), pp. 75–84.

UITP - INTERNATIONAL ASSOCIATION OF PUBLIC TRANSPORT (n.d.) *Statistics*. Available from: http://www.uitp.org/knowledge/Statistics.cfm [Accessed 04/10/2010].

UITP – INTERNATIONAL ASSOCIATION OF PUBLIC TRANSPORT (2009) *Assessing the benefits of public transport*. Available from: http://www.uitp.org/mos/focus/FPBenefits-en.pdf [Accessed 06/04/2010].

URRY, J. (2000) Sociology beyond societies. Mobilities for the twenty-first century. London: Routledge.

URRY, J. (2004) The 'system' of automobility. *Theory, Culture & Society*, 21 (4/5), pp. 25–39.

URRY, J. (2006) Inhabiting the car. *Sociological Review*, 54 (s1), pp. 17–31.

WATSON, S. (2006) City publics. The (dis)enchantments of urban encounters. New York: Routledge.

CHAPTER 2

Architectural Visualizations as Promoters of Urban Aestheticization. A Visual Culture Approach

Tobias Scheidegger

Introduction

In a publication about *Epcot Center*, a section of Disney World in Florida, I found several images dating from the 1960s, when the center was still in the planning stage (Beard, 1982). Those pictures perfectly illustrate the topic of my essay. In the first photo we see Walt Disney himself, standing in the middle of an undeveloped area, holding a big map – "studying the blueprint for the future" as it's described in the title.

Another photo shows the busy entertainment entrepreneur sitting in a studio, surrounded by cameras and presenting a copy of said blueprints especially designed for the public. This series is completed by a colorful painting which over a double page portrays the yet unbuilt *Epcot Center* as an astonishing heaven on earth: using a visual language which combines the artwork of the Jehovah's Witnesses' *Watch Tower* with a certain socialist realism, this luminous visualization articulates the promise of a bright future coming along with this building project.

Why am I discussing these pictures? I'm not specifically interested in the engineering of leisure parks or in the self-expression of Walt Disney. Rather my essay deals with images of construction projects as specific forms of representation of urban (public) space. However, starting with images produced by Disney, one of the world's largest entertainment companies, is not merely accidental. The use of a visual language of advertisement which portrays space as a materialization of commodified dreams characterizes visualizations of not only utopian pleasure grounds, but also other building projects which on the surface seem to be much more reputable: opera houses, libraries, museums, and corporate headquarters.

The contemporary genre of these architectural visualizations, on which the following essay focuses, is not as fancy as in the nineteen-sixties anymore, but is nevertheless quite interesting for researchers on transformations of urban public space. In the studios of Walt Disney, as in most architectural offices, computers have replaced the drawing tables; today building projects are commonly planned and, first and foremost, mediated by computer-generated architectural visualizations (also called "renderings"[30]). In our everyday life, we regularly meet with those renderings. In newspapers, on webpages or billboards, those colorful pictures provide information about buildings that are yet in the planning stage. But, as I want to outline, those images are not just communicating the physical dimensions of a construction project, but also certain ideas and concepts about urban public space and its reshape; that is to say, they depict a political version of Disney's "blueprint for the future" quoted above. The two guiding themes of this book in particular – mobility and aestheticization – are prominently illustrated in the visualizations. In this article, I'll specifically demonstrate the way architectural visualizations display as well as promote tendencies of aestheticization of urban public space.

Visualizations as a means for image control

We know much about the use of visual means and media throughout the recent history of architecture. One finds a lot of books dealing with the emergence of plans and their perspective representations in the princely urbanism of the Renaissance and Baroque, as well as various works on the use of film and photography for the propagandization of totalitarian architecture (see e.g. Lefebvre, 2003 or Schönberger, 1988). Less attention, however, has so far been directed toward the media for the communication of contemporary architecture (for exceptions see e.g. Jabusch, 1997 or Söderström, 2000). Today's developments are mediated by computer-generated renderings, in a visual format that shall be shortly outlined. First I want to explore some theoretical aspects of this visual genre and lay out evidence of how the interplay between visualization and the resulting architecture participates in the creation of urban reality. I aim to prove that computer-generated visualizations show, as did their predecessors, more than just information about a planned building – they transport specific world views and concepts of mankind[31].

The professional discussion about the use of renderings in architecture communications is meager. In interviews with image producers, the same reasoning for the legitimization of this genre are repeatedly put forward: renderings are considered as being less abstract than plans and therefore can be "read" by laymen, thus contributing to the democratization of communication of

---

30 Rendering: virtual computational representation in three dimensions of objects in development. For the terminological history of »rendering« see Oechslin (1987).

31 This property is not a pure distinguishing mark of digital renderings. As several authors point out already, older forms of architectural renderings are used to transport more then just mere information on a planned building but always some kind of ideology. (See e.g. Müller, 1993, p.65).

and on architecture[32]. This professionals' point of view is limited and will be in some ways contradicted with good cause in this article.

At first sight the manipulative potential of rendering is evident. 'Suggestion instead of information' seems to be the formula. In the first place, the varied aesthetic loading of visualizations – solemn or even 'sacral' light effects such as sunsets, shiny façades, etc. – induces an emotionalization of the image contents. Such images inhibit the onlookers from investing their reflexive capacities, addressing them specifically from their sensualist side. The elaborate finish, a more or less perfectionist photorealistic image[33], gives it the character of "significative authority". With this concept, Swiss geographer Ola Söderström describes the convincing power of the visual in collective processes of urban planning: the visual has the potentiality of mobilizing and unifying different players involved in controversial negotiations on planning and building (Söderström, 2000).

These properties push the issue of democratization to the point of absurdity. Professional photorealistic renderings make investors' dreams appear as anticipated, already-built realities and attempt to act as self-fulfilling prophecies. Various producers stress the fact that, through such a visual operation, the promoters of real estate development gain an important strategic advantage. The discursive battlefield for potential enmity is withdrawn by the sheer self-evidence of the realistic image, while arguments about sense and aim of architectural intervention are relocated to the level of the purely aesthetic—a matter of bigger or smaller cubature, of more or less glossy materials. These parameters lead inescapably to the conclusion that the genre must be identified as a tool for the control of visual imagery, aiming at exerting influence beyond the individual building and onto the urban imaginary[34].

---

32 Four qualitative, guideline-based interviews in 2006 with visualization designers based in Zurich were part of the empirical material for my master thesis on the visual culture of architectural renderings. For more details see Scheidegger, 2009.

33 For further discussion on the entanglement of photo-realism and visualizations see Jabusch 1997.

34 There's a close affinity between contemporary architecture with its iconic buildings aiming to gain impact on cities' identities (Hubbard, 1996) and the visualizations as transmitters of this identity policy. The primacy of mediatization since the 1980s has had enduring effects on contemporary architecture, especially concerning the role of surfaces (Müller and Dröge, 2005, p.15).

An ethnography of digital urbanity

Having sketched some theoretical aspects of the visual genre of architectural renderings, I would like to start a short ethnographic journey into this virtual urban space. The starting point of this *dérive* through a city of bytes is the assumption that the visualizations can be read as materializations of specific spatial conceptions common among professional image producers. These image producers, such as architects, designers, or information scientists, belong to a heterogeneous social group defined by German art theorists Michael Müller and Franz Dröge (2005) as "symbol producers". With this term they describe the contemporary urban elite of creative professionals, which is highly involved in the processes of transformation of urban space. One aspect which is especially highlighted by Müller and Dröge (2005) is the profound connection between changes in hegemonic urban lifestyles and processes of reshaping urban space. This focus on the *combination* of changes in lifestyle and spatial concepts is a useful specification for the analysis of transformations of (urban) space. The theorists describe this interconnection in the following way:

> Considering those indications, it's quite obvious to assume – as is already known for lifestyles – that also the conceptions of space aren't just changing, interfering or disappearing. The creation of these spatial concepts are experimental grounds for the professional elite of today's middle-class, which are fed back in milieus constituting urbanistic discourses. This system-stabilizing feedback effect leads us to the conclusion that over the next years some few dominant combinations of lifestyles and spatial conceptions will gain hegemony in this experimental ground and shape the design of inner cities as well as the design of some rural areas. (Müller and Dröge, 2005, pp. 92–93, transl. by T.S.)

Following Müller and Dröge, it could be argued that the conceptions of space communicated and promoted in architectural visualizations are strongly interrelated with certain lifestyles. This means that those pictures can be read as materializations of those combinations of lifestyle and spatial conceptions.

A concrete case shall exemplify my diagnosis. My example is taken from two different visions of the Limmatplatz in Zurich *(see Pictures 1 & 2)*. This square is situated in a former working-class neighborhood which has been transformed since the 1990s by a general gentrification. In 2007 the central tram stop which gives the square its function and character was rebuilt. Two images – a photograph I took of the ancient stop and the visualization of the planned new stop – give evidence of the anticipative character of visualizations. The comparison of 'before and after', a visual structure well known from advertisements for weight-reduction, etc., not only informs viewers about the planned measurements and appearance of the tram stop, but also makes a statement about the social implications of the new building.

Picture 1: Everyday life at the former tram station at Zurich's Limmatplatz. Photograph by Tobias Scheidegger

Picture 2: Mobilizing physical and social space: visualization of the planned tram station. Photograph from billboard by Tobias Scheidegger

On the photo of the ancient square, small groups of different kinds of people squat on the benches. The walls behind them are tablets, on which tags and patina testify to many layers of appropriation. The 'new Limmatplatz' will be different; such is the promise of the information boards which were erected all over the square by the town authorities around the same time when my first photograph was taken: it appears as a repressive Feng-Shui in concrete, a masterpiece of panopticism. This strategic intention was openly confirmed by the town's project leader in an interview with a journalist, pointing out its advantages: the new tram stop would be "enjoyable for a stopover of several minutes, but not of several hours" (Meier, 2004, transl. by T.S.). It allows views from all sides and has no backside like the old building. The naked benches in the rendering are empty and the former users have disappeared. Instead, small groups of energetic individuals on the move – symbol producers, one could assume – can be seen.

Reshaped space I: Mobility and control of space

One of the most common characteristics of public space as portrayed on visualizations is its inherently dynamic nature; the renderings speak a language of mobility. People crowding the images are mostly on the move, while city dwellers sitting or hanging around are a rare exception. We behold people at rest only in such cases where this kind of use is explicitly required by the architectural project, as in recreation areas and such.

In accordance with film theorist James Donald, I suggest that this kind of primacy of mobility goes hand in hand with a certain devaluation of urban public space. In reference to Hillis Miller, Donald argues that space is always constituted through an act of *taking place:*

> Almost paradoxically, he suggests that we cannot imagine space as such. What we imagine, he suggests, is always an event or events taking place. Our imagination is inherently narrative. Space is less the already existing setting for such stories, than the constitution of space through that *taking place*, through the act of narration. (Donald, 1999, p. 123)

In this world of images, one observes a severe lack of *taking place* accompanied by a harsh neutralization of the narrative character of space. Reasons for this process of neutralization may be many but it's quite obvious that urban fears, real as well as imaginary ones (see McElligott, 2005), are strengthening the tendencies that underlie them. *Taking place* events always have the potentiality of unpleasantness occurring, therefore in times of zero tolerance, consensus tends toward the preference for uneventful public space by means of mobilization, safe distances, and control.

This consensus finds itself materialized in the visualizations. One of my interview partners, a professional producer of architectural visualizations, described an essential rule for the visual language of safety, which is highly appreciated by his customers: "Never dark corners – fear of a crime is something which is never allowed to happen. We strongly recognize our customers' need for safety."[35]

The incarnated paradigm for this kind of use of space, characterized by the primacy of sterile mobility, is the energetic white collar worker. Often with a cell phone on his ear and carrying a briefcase, he energetically crosses the virtual urban space. He (and not she, by the way, at least for the vast majority of my image samples) has become a symbol used in almost all visualizations, which leads me to my second point about the characteristics of public space represented in architectural visualizations.

Reshaped space II: Aestheticization&normativity

The visual primacy of quasi-Simmelian flow and mobility, as described above, equates to the primacy of economy. People in the visualizations seem to be digital materializations of post-Fordist self-management: even if they stand still for a short chat, they seem to be doing business. As already mentioned, this ideology is visually symbolized by the white collar worker who appears on quite every image. The message is obvious; he is the embodied promise of economic prosperity that comes along with the new building. But the visualizations of urban public space are inhabited not only by white collar workers. There's a slight diversity of city dwellers represented on the images, as one of my interview partners, who works as image producer, pointed out: "We have the task to picture two kinds of people, that is to say, to differentiate between business people and casual people who are just living there, doing some leisure activities. These different groups are two quantities which have to be balanced well."[36]

Those two separated spheres of middle-class everyday life – job and leisure – are held together by an overall poetics of consumption, the language of a meaningful aestheticization of public space and its users. The public space of

---

35 Interview with a visualization specialist by Tobias Scheidegger in Zurich (2006), for more details see Scheidegger 2009.
36 Interview with a visualization specialist by Tobias Scheidegger in Zurich (2006), for more details see Scheidegger 2009.

ArCAADia (see Scheidegger, 2009) is dominated by elegant suits and trendy bags. As confirmed by my interview partners – mainly architects and image producers – this aestheticization is not at all insignificant or even accidental. Talking about visualizations of apartments, one of the image producers mentioned the so-called "rule of 20 percent", which means that the furniture pictured in those visualizations is usually about 20 percent costlier than the furniture that the consumer addressed by the visualization can afford in reality. This same visual language of over-styling characterizes the visualizations of urban public space, including city dwellers and their clothing, but beyond that, even cars, trees and so on are chosen by criteria of aesthetics and social prestige.

Thus visualizations aggressively promoting a normative aestheticization of urban public space go hand in hand with the economical or policed exclusion of people who don't fit into the narrowly defined space of business and leisure of the middle class (see e.g. Ronneberger, 2001). This dominance of an aestheticized public tends to create a visual and spatial tautology: because public space is predominantly used for economic purposes, there are little or no people in the visualizations who are not doing business or shopping. Those people not in accordance with the dominant logic of the use of space are therefore not represented in the image. This disappearance is not merely virtual, however; invisibility in the images tends to become strived-for reality also in real space, which leads me to my last point.

Spatio-functional determinism

In the light of its genuinely commercial character, it's not astonishing that some categories of people who intensively use public space, such as drug addicts or the homeless, are not represented in these visualizations. But the analysis of a sample of images reveals that surprisingly enough, several other more legitimized categories of city habitants are likewise invisible in the images, mostly elderly people and children, including teenagers. Let me briefly focus on the youth.

While – at least in the visualizations – urban public space is increasingly shaped as an aestheticized stage for business and leisure, youth somehow don't fit into this world of adult activities[37]. The only remaining areas for kids in the visualizations are those places deliberately designed primarily for children's play or leisure activities, such as parks and so on. Thus urban public spaces as portrayed in these images bear witness to the processes of *"Verhäuslichung"*

---

37 For recent researches on contemporary exclusions of children and teenagers from urban public space see Muri and Friedrich (2009).

(domestication) of the youth, as described by German social pedagogue Jürgen Zinnecker. During the last few decades, juvenile outdoor activities have been steadily banned to the indoors for several reasons. First and foremost are the severe impacts of the shaping of urban space by individual motor car traffic since the 1960s[38]. One can move from there into many other explanations, including the increasing temporal structuring of juvenile everyday life or the increasing preference for electronic devices like TV or computer games. One important reason may also be found in changing attitudes toward the use of public space itself. The more that space surrounding public or corporate buildings is conceived as representative space, the more that adolescents using this space for activities like skateboarding or just hanging around become perceived as a potential danger to the narrowly-defined specific function of the space. Driven by the fear of littering, noise or vandalism, an adult public is more readily willing to impose security guards and prohibitive rules in order to ensure the space will be undisturbed as necessitated by the serious matters of its function. As already described in regards to the marginalized city dwellers made invisible in the renderings, the politics of zero tolerance goes hand in hand with the politics of visualization: out of sight, out of mind.

But in this world of contemporary architectural visualizations, not only the kids are banished to green areas. As indicated above, asphalt and sidewalks function here as metaphors of the space for business. With its predominantly male public imagery, this space is quite explicitly a gendered one. Gangs of white-collar workers with briefcases distinctly mark the visual field of attention. Of course women aren't completely absent in the images of busy areas in front of business premises, but their presence – often characterized as carrying out a leisure activity like shopping (i.e. the second mode of existence provided by the creators of renderings) – does not belie a certain male primacy of those spaces. Another spatial counterpart conceded to female city dwellers can be found in the parks. Visualizations of green areas are the ones allowed to be enriched with signs of femininity and familiarity: mothers driving buggies or chatting in the shadows of trees while their children enjoy unrestricted possibilities to frolic around. Constituting a space of semi-privacy within the specific urban public space as shaped in this visual genre of renderings, those pictures give evidence of the devaluation of urban public space, accompanied by a spatial segregation which reflects the fragmentation of everyday life in developed capitalism. It is for this reason that these gendered green open spaces were harshly criticized by Henri Lefebvre in the 1960s as pale imitations both of nature and of ideal public space:

---

38   For this subject see the article by Giulio Mattioli in this book.

What no longer has a meaning is given one through the mediation of a fetishistic world of nature. Undiscoverable, fugitive, ravaged, the residue of urbanization and industrialization, nature can be found everywhere, from femininity to the most mundane object. Parks and open spaces, the last word in good intentions and bad urban representation, are simply a poor substitute for nature, the degraded simulacrum of the open space characteristic of encounters, games, parks, gardens and public squares. (Lefebvre, 2003, p. 27)

The simulation of urban public and community

Whether one agrees with Lefebvre's analysis under the banner of alienation or not, it's quite obvious that many architectural renderings tend to portray spaces which can plausibly be interpreted as simulacra, highly functioning as compensatory spaces. For that reason the foregoing characterization of the urban public space depicted on the renderings as clean and neutralized from every narrativity has to be complemented. There is in fact a category of images, which functions quite the opposite way– by anxiously evoking narratives of community and vitality. As in built reality and in the renderings, the construction of controlled space defined by guidelines of visibility and transparency involves the potential danger of producing a dead urban desert. Often the results of exaggerated spatial cleanliness undermine the very intention of such architecture: feelings of safety and belonging. So at least in the visualizations, architects and city planners have the possibility to project ideals of a livable 'urbanity' on those sterile places. This visual strategy of artificial animation was also confirmed by one of my interview partners, who often produces renderings for urban planning in Zurich:

> Well...there are those fears, that it ain't lively enough. This is more or less the principal thing. There is a consensus concerning the importance of animation. That's what the town fathers want: there are children, it's bright, it's lively, it's as well lively in the evenings when it's dark and women don't have fear to walk by and so on. Because when it's dead and there's a lack of animation, then it's quite problematic.[39]

What can be observed by having a closer look on this genre of animating visualizations? Those pictures display a public space, which is characterized by a general heterogeneity; by diversity of race, age and gender (not necessarily by class), by pluralistic use of space such as sitting, walking, climbing and jumping and last by a richness of activities such as work or play, all of them properties which contradict the functional determinism of space as usually experienced in

---

39 Interview with a rendering specialist by Tobias Scheidegger in Zurich (2006), for more details see Scheidegger 2009.

everyday life in European cities. This kind of Potemkin animation is common not only among visualizations of planned public buildings or places, but also and even more in renderings of commercial objects such as shopping malls, urban entertainment centers or football stadiums. Despite the fact that the individual liberty of action in those areas is quite often restricted by pedantic site rules and security services, these images depict scenes of lively, playful and even adventurous spaces – as if unexpectedly Guy Debord had become the chief executive officer of the portrayed shopping mall. This visual language of simulated animation equates to the self-conception of those private commercial spaces, which advertise themselves as a fully adequate replacement of the ideal city. German cultural theorist Klaus Ronneberger describes these semantics of substitution as follows:

> Such archipelagos of controlled experiences try to manufacture the atmosphere and the image of the traditional city square, which commonly is compared with communication, publicness and spectacle. (Ronneberger, 2001, p. 31, transl. by T.S.)

In spite of the fact that this *mise-en-scène* still echoes a slight idea of the utopian (ethical and political) conceptions of 'urbanity', the very nature of those stages of controlled experience is thinly disguised. This simulacra of urban ideals is mostly nothing more then a poor attempt to maximize customer's resting time in a space of commerce.

Although this reading of the simulation of urban public in architectural visualizations as sheer mantling of commerce is reasonable, there are fair objections that contradict this critical interpretation. To come full circle, the visual world of Walt Disney shall take the center stage again. In an article about the so-called "Disney Syndrome', Austrian architecture critic Robert Schediwy reviews common complaints against postmodern architecture – with Disneyland as its materialized Mecca – which are pervasive amongst professional architects, critics, professors and so on. Schediwy provocatively points out that those professional reservations against Disney architecture and its simulating/faking character are not shared by the public at large. Quite the contrary, the preferences for such an idyll in plastic and concrete is interpreted by him as a popular architecture critique against the failed ideals of modern architecture:

> Gradually also the most elite spirits start to recognize, that Disney architecture and city planning, despite of and because of its nostalgic components answer human needs better than the standard bearers of modernity ever succeeded. (Schediwy, 2005, p. 333, transl. by T.S.).

Those two separate spheres of middle-class everyday life – job and leisure – are held together by an overall poetics of consumption

Totally conscious of the socially-segregating nature of such Disney ideals brought into being – as in the example of "Celebration", a gated community for 20,000 residents realized by the Disney company in Florida – Schediwy advocates for dealing seriously with the reasons behind the popularity of such architecture. As is plausible for this kind of architecture, visualizations simulating highly animated urban public space can also be read as reflections of deep-rooted popular desire. Maybe it's not at all accidental that in one of my interviews a professional producer of architectural visualizations declared that he is manufacturing "fairy tale worlds". In the same way the worldwide success of Disney's fairy tale movies can't be explained away just by a false consciousness affinity for kitsch, the portrayal of idealized scenes of urban public life in those specific renderings – with people gathering, chatting or playing and, last but not least, a remarkable absence of car traffic – is not merely a commercial promotion but likewise reflects longings for public spaces worth living in.

Summary/Conclusion

I tried to read architectural visualizations not just as a means of communicating physical dimensions of a building project yet in the planning stage, but as a materialization of a combination of lifestyle and spatial conceptions. Thus public space represented by and for middle-class symbol producers is a space whose narrative nature – and therefore the potential danger of (pleasant and unpleasant) events happening – is banished by spatial control and mobilization. The public crowding this cleansed space is visually reproducing a normative aesthetic of business and leisure. 'Classic' urban public space of the inner cities like streets, squares and sidewalks is predominantly portrayed as an adult and male space. Other groups of city dwellers that do not fit into the narrow determinism of the possible ways of using this space are made invisible on the visualizations. While some categories of the population are totally erased from the visual world of renderings, others are banished to narrowly defined spatial sectors; e.g. the youth and the mothers, which 'have to' inhabit the green areas.

With their emphasis on clearly defined, clean and transparent spaces, both the visualizations and the built architecture run the risk of producing deserted spaces, a danger which is averted in the visualization by means of artificial animation. Some categories of visualizations thus simulate idealized images of crowded and multi-functional urban public space.

This analysis revealed the ongoing oscillation between the visualizations and built space, and that the two can never be entirely separated. This leads to a conclusion that architectural visualizations are potential tools for the "imagineering" (Tom Holert, 2000) of space that are not only representing an overall reshaping of urban public space, but also effectively *promoting* such changes.

References

BEARD, R. (1982): Walt Disney's EPCOT Center. Creating the New World of Tomorrow. New York: Harry N. Abrams.
DONALD, J. (1999): Imagining the Modern City. London: AthlonePress.
HOLERT, T. (2000): Bildfähigkeiten. Visuelle Kultur, Repräsentationskritik und Politik der Sichtbarkeit. In: HOLERT, T. (ed.): Imagineering. Visuelle Kultur und Politik der Sichtbarkeit. Köln: Oktagon-Verlag, pp. 14–33.
HUBBARD, P. (1996): Urban Design and City Regeneration: Social Representations of Entrepreneurial Landscapes. Urban Studies. An

International Journal for Research in Urban and Regional Studies, 33 (8), pp. 1441–1461.
JABUSCH, D. (1997): Das digitale Bild der Stadt. Göttingen: Cuvillier.
LEFEBVRE, H. (2003): The urban revolution. Transl. by Robert Bononno; foreword by Neil Smith. Minneapolis: University of Minnesota Press.
MCELLIGOTT, A. (2005): Urban Fears and ‚Virtual Cities': Dystopian Perceptions/Utopian Environments. IMS. Informationen zur modernen Stadtgeschichte, (1), pp. 59–74.
MEIER, J. (2004): Mehr als nur eine Tramhaltestelle. Neue Zürcher Zeitung, 20th Oct, p. 53.
MÜLLER, A. M. (1993): Aber wessen das Gefäß ist gefüllt, davon es sprudelt und überquillt. In: MUSEUM FÜR GESTALTUNG ZÜRICH (ed.) New Realities – Neue Wirklichkeiten. Entwerfen im Zeichen des Computers. Zürich: Museum für Gestaltung, pp. 45–67.
MÜLLER, M. and DRÖGE F. (2005): Die ausgestellte Stadt. Zur Differenz von Ort und Raum. Basel und Gütersloh: Birkhäuser.
MURI, G. and FRIEDRICH, S. (2009): Stadt(t)räume – Alltagsräume? Jugendkulturen zwischen geplanter und gelebter Urbanität. Wiesbaden: VS Verlag für Sozialwissenschaften.
OECHSLIN, W. (1987): Die verführerische Zeichnung - seductive drawing, DAIDALOS, 25.
RONNEBERGER, K. (2001): Konsumfestungen und Raumpatrouillen. Der Ausbau der Städte zu Erlebnislandschaften. In: BECKER, J. (ed.): Bignes? Size does matter. Image/Politik. Städtisches Handeln. Kritik der unternehmerischen Stadt. Berlin: Stephan Greene, pp. 28-41.
SCHEDIWY, R. (2005): Postmoderne. Alles Kulisse? – Das Disney-Syndrom. In: SCHEDIWY, R.: Städtebilder. Reflexionen zum Wandel in Architektur und Urbanistik. 2nd ed. Wien: Lit, pp. 331-334.
SCHEIDEGGER, T. (2009): Flanieren in ArCAADia. Digitale Architekturvisualisierungen – Analyse einer unbeachteten Bildgattung. Zürich: Institut für Populäre Kulturen.
SCHÖNBERGER, A. (1988): Architekturmodelle zwischen Illusion und Simulation. In: SCHÖNBERGER, A., INTERNAT. DESIGN ZENTRUM BERLIN (eds.): Simulation und Wirklichkeit. Design, Architektur, Film, Naturwissenschaft, Ökologie, Ökonomie, Psychologie. Berlin: DuMont, pp. 41–54.
SÖDERSTRÖM, O. (2000): Des images pour agir. Le visuel en urbanisme. Lausanne: Editions Payot.

CHAPTER 3

Public Art Spaces: The Dilemma of Economic Growth and Social Inclusion in an Aesthetised Urban Context. A Case Study of Initiative "Intermediae", Matadero Madrid, Spain

Clara Fohrbeck

Introduction

The Mediterranean City Type (Leontidou, 2003)[40] clearly presents to its visitors that feasible urban public space – despite the rapid development of a 'network society' in the course of the 'information age' – has not lost all its relevance (Castells 2000). In Madrid, one finds little plazas everywhere, where some sit on benches discussing their daily affairs and parents watch their children play amid the chaos of the inner city centre. The Internet certainly has not succeeded in tying Madrid's inhabitants to their homes, hindering them from seeking their social contacts, enjoying a fresh breeze on the streets or strolling around traditional market halls. The city centre of Madrid represents a space where tourists, illegal immigrants and inhabitants mingle, unconsciously forming a part of the city's heterogeneous imagery. However, the heterogeneity does not mean equality within public space. Rather, "there is always great competition over its control. Whoever controls public space sets the "program" for representing society" (Zukin, 1998 b).

In democratic society, 'equal access' is considered a value. This is why whoever controls the 'public good' must at least pretend to be providing opportunities for equal access to it for all members of society. Public space in the city is not an exception in this concern, and free access to it for the city public is considered both a value and a rule. The question to be explored is how this rule actually functions within the reality of our cities. The 'inclusion of a broad public' is presented as one of the main objectives of the initiative 'Intermediae', which is part of the project of the 'Centre of Contemporary Creation' 'Matadero Madrid'[41]. This enormous project, subsidised with 111,000,000 Euros (of which 75% are paid by the city), is being developed and realised on the grounds of an old industrial site that is situated on the peripheries

---

40 In her paper represented at the symposium "(In)visible Cities. Spaces of Hope, Spaces of Citizenship". Centre of Contemporary Culture of Barcelona, 25-27 July 2003, Lila Leontidou compares Mediterranean representations of cities as Spaces of Citizenship vs. British representations as Spaces of Risk.

41 'Matadero' translated into English means slaughterhouse

of Madrid's city centre. There are various facets of the urban context that play a crucial role in the development of a cultural project like the Matadero Madrid. Madrid is engaged in the processes of urban regeneration. The district of Arganzuela, where Matadero Madrid is located, is one such site. The district provides fertile ground for numerous matters of social deprivation and social cleavages. It also provides abundant economic motivation for the government to attempt to extend the size of a globally and aesthetically attractive city centre through the transformation of this slaughterhouse. Although Arganzuela's industrial activity is a thing of the past, its legacy is still reflected by its relatively poor housing conditions and a continued tradition of work and workers. The area's rapid transformations have resulted in a lack of social infrastructure, as it proved unable to develop simultaneously from an industrial territory into a residential area (Sanchez, 2008, p. 9). At first glance it might seem paradoxical that a prestigious centre for art, design, theatre, architecture and other areas of 'contemporary creation' is being developed here. However, according to Miles, "given that culturally-led redevelopment occurs in de-industrialised conditions, it is not surprising that outposts of cultural recoding are geographically juxtaposed with areas of residual deprivation" (Miles, 2005, p. 893).

Still, implementing cultural infrastructure into the physiognomy of cities does not necessarily result in interaction between different public spheres. The spread of post-Fordist conditions bears with them new manners of social regulation that are based on a changing form of urban governance. These social modes of regulation might function on a short-term basis but, in the long run, there is a tendency towards the reverse effect, one of social marginalisation within regenerated districts, as Mayer notes. The focus on aesthetised urban regeneration projects and its dominance over social objectives is thus inclined to create cleavages within society and runs the danger of undermining the concept of a civil society (Mayer, 1994, p. 332). This paper will address the relation between art and public within a city context. I will discuss the opportunities and challenges in providing access to contemporary cultural and art activities for the broader public in the course of implementing aesthetised urban regeneration projects. I will try to elaborate on this issue by posing the following question: To what extent can "public-ness" exist in an art centre in which the discourse of contemporary art seems predestined to exclude the non-dominant publics, such as the immigrant population of the neighbourhood?

First of all, I will briefly define the term 'public' in order to be clear on how this paper will analyse the challenges of creating a platform where a broad 'public' can gather and interact. Then, two contrasting functions of art – economic and social – will be elaborated in order to understand the role art plays when city life has become increasingly aesthetised. With this theoretical framework in place, the rationale of establishing a centre like the 'Matadero Madrid' in the context of city planning can be explored. The initiative 'Intermediae' takes on the responsibility of opening up this centre to a broader public. Challenges remaining and opportunities arising from the creation of a platform for 'inter-public relations' will be examined by discussing the way in which artists and the initiative have interacted with locals in order to achieve their goal of creating a 'public' art space.

Public space

In order to understand whether open public access to 'Matadero Madrid' centre is provided, it is necessary to start by defining the term 'public'. This work of definition will indicate the challenges that a gap between theory and practice of urban public life might bear in the context of creating aesthetised public urban space. In very general terms, the 'public' represents a platform where different people with different public interests may and do interact.

One of the most influential writings on the concept of a 'public sphere' is Habermas' classical work 'The Structural Transformation of the Public Sphere - An inquiry into a Category of Bourgeois Society', in which he defines the term as follows:

> the realm of our social life in which something approaching public opinion can be formed [...]. Citizens behave as a public body when they confer in an unrestricted fashion - that is, with the guarantee of freedom of assembly and association and the freedom to express and publish their opinions – about matters of general interest [...]. The expression 'public opinion' refers to the tasks of criticism and control which a public body of citizens informally practices [...] vis-a-vis a ruling class. (Habermas, 1989 quoted in Pusey, 1987, p. 89)

According to the Habermasian definition, Sharon Zukin (1995) described one of the core conditions for public space as the possibility to be "used by many people for common purpose". Who are these 'many' people though, and to what extent is it possible to create a space where many people have a 'common purpose' to use it? As long as 'many people' are heterogeneous and follow different public interests – the essence, in a way, of a heterogeneous urban environment – a public space is likely one that should not only be used by many people with one common purpose, but also by many different people with different purposes.

In this paper, the 'public' will be perceived in the light of Fraser's (1992) approach to the topic of public spheres in her work 'Rethinking the Public

Sphere'. Fraser suggests that Habermas' definition of the bourgeois-liberal public represents a place of exclusion as such. She defines the official public as the generally-accepted discourse of the dominant power, since it takes on the role of so-called 'public interest' and therewith defines concepts and perspectives for all other subordinated publics. Such a public does not include all citizens of a city. However, 'alternative publics', which Fraser defines as 'subaltern counter-publics', may be generated. Those 'subaltern counter-publics' are parallel discursive realms whose members discuss with each other and are generally opposed to opinions, concepts and perspectives of the dominating public. Fraser distinguishes between strong and weak publics; the latter accordingly have difficulties with access to the dominant position in the sphere where mainstream public opinion is formulated. With her claim for 'inter-public relations', Fraser sees the necessity in establishing an open exchange between the different publics in order to enable access to a certain set of problems, terms and definitions of the dominant discourse. Only under these conditions may the expectation of creating a free public debate and an 'open' public be fulfilled (Fraser, 1992).

Is it then possible for a centre of contemporary creation to serve as "a meeting point for professionals from the world of culture, for creators and for the public in general"[42] – which is one of the ten objectives that the Matadero claims to follow? What kinds of barriers may appear when professionals, artists and the 'public in general' are supposed to interact inside the premises of the initiative 'Intermediae'?

Dimensions of artistic creation

To be able to approach this question, it is crucial to elaborate two contrasting functions of art which evolved out of professional and public discourses about the new role of art and culture in the 20th century within the context of urban regeneration: economic growth and social inclusion.

The increasing economic role of art and culture in late modern urbanisation

The deindustrialisation process has triggered major economic transformations. In the early 20th century, the classical factory towns and cities dominated 'Western' capitalist countries. Fordist mass production was perceived to be a typical characteristic of the evolving industrial metropolis. However, a transition away from mass production and clearly-assigned labour structures to an economy of post-Fordism occurred, characterised by the strategy of 'flexible

---
42 Named as one of the ten objectives on Matadero's official website in 2009.

accumulation'. This brought with it an overall new manner of urbanization (Scott, 2006, p. 7). According to Mayer,

> as a consequence of this economic development, diverse efforts to mobilise and coordinate local potential for economic growth together have produced the effect of gradually undermining the traditional sharp distinctions between different policy areas [...and] educational, environmental and cultural policies have become more integrated with, and are often part and parcel of economic development measures (Mayer, 1994, p. 327).

A physical result of the industrial era's decay are the many vast available spaces created due to the strong denunciation of industrial activity – including the premises of the slaughterhouse in Madrid. It has become a mission for many urban planners to find solutions for how to transform old industrial sites through aestheticisation. This is realised within the context of so-called 'urban regeneration' – projects that often represent the effort to establish innovative economic parameters in order to revitalise abandoned parts within city centres. Urban planning in the last few decades has been affected by an increasing global economic competitiveness. Part of contemporary urbanisation is the process by which 'individual' cities get constituted as systems of internal transactions embedded in a wider system of transactions binding all cities together into a grid of complementary and competitive relationships (Scott, 2006, p. 5).

In the post-modern era, new sectors have evolved and become the means of economic growth and innovation: "high-technology industries, neo-artisan manufacturing business and financial services, cultural-products industries" (Scott, 2006, p. 7). One can detect the relevant impact of cultural industries on the economy of Spain. According to annual statistics, the cultural sector is characterised by continuous growth in the time period from 2000–2008 at a yearly average rate of 6.7% of GDP. This rate is related to growth in the initiatives of entrepreneurship: 70,109 new cultural enterprises emerged in 2008, generating an employment rate that is high above the average of the Spanish economy. Of the cultural enterprises throughout Spain, the majority are centered in Madrid (25.4%) (Frende Vega & Pena Sanchez, 2010). Since the cultural industries sector has become increasingly significant to cities willing to compete on a global scale, one can argue that this provides endless opportunities for "bringing the dimensions of economy, culture and place into some sort of practical and humanely responsible harmony" (Scott, 2006, p. 7).

Zukin (1998) emphasizes the increase in the presentation of symbolic qualities of places, which in turn depicts an important strategy of raising land value. Cities or city quarters succeed in epitomizing societal aspects if cultural values are transformed into a visual image. A further relevant aspect that Zukin mentions about cultural sites and industries is the growth of a network of communication and services between local actors that evolves out of context-related knowledge. This can only be optimized within systems that are locally integrated. Cultural knowledge and its exchange form a great part in this kind of networking and emphasises the relevance of local conditions and experiences. Therewith Zukin argues that the local is not eliminated by the global, but instead fosters interest in the heterogeneity of the local in a world becoming increasingly homogeneous.

Art as a tool for social inclusion

At the same time, social, cultural and economic exclusion of all types is constituting new parameters of social (in)equality within urban landscapes. Income distribution, in this case, is not always an indicator of a socially-inclusive urbanisation. Moreover, when looking at art in this context, one should take into account the cornerstones of democracy and citizenship and the integration of various social dimensions into a vibrant life of the city "not just for its own sake but also as a means of giving free reign to the creative powers of the citizen at large" (Scott, 2006, p. 36) – and therewith, in the terms of Fraser (1992), creating sites of 'inter-public relations'. Art is increasingly perceived as an instrument that works against economic domination in the cultural sector and fosters the social inclusion of local communities, in order to provide a bridge between global and local networks:
 The task is not to create an 'alternative sector', but to make it accessible to and resourceful for marginalised groups threatened by the powerful polarisation processes of post-Fordism (Mayer, 1994, p. 331). The key to the creation of social cohesion is:

> the belief that public art, or the processes through which it is produced, is able to generate a sense of ownership forging the connection between citizens, city spaces and their meaning as places through which subjectivity is constructed (Sharp, Pollock and Paddison, 2005, p. 1003).

Williams (1997), supports this view and suggests focusing on group cooperation when it comes to artistic production in order to define the common interests of the community and therewith establish a common identity. The competitive aspect of art should be replaced by collaboration between artists and the public as should artistic self-interest by group needs. Concerning the positive effects of social inclusion through art on community development, the Community Development Foundation (2001) points out that the arts are a crucial means of revitalising 'economically, socially and culturally disadvantaged areas' in order to foster the development of urban communities. Participation within the arts, according to an arts impact study of Williams (1997), have demonstrated to be highly effective in strengthening factors that indicate an increase in community development, such as a strengthening of local identities, an increase of social integration and the provision of recreational activities. Blake Stevenson Ltd. (2000) carried out case studies to measure the impact of art projects on the sustainable regeneration of marginalised city districts and the fostering of social inclusion. Various effects were indicated, including the improvement of individual personal skills, an improved external and internal impression of the district and the fostering of economic investment, again showing the interrelation between art as a tool of economic growth and as an instrument of social inclusion.

The richness of case studies concerning the impact of art on community development insist that art, besides serving as a tool for economic competitiveness on the global scale, can also play a crucial role in the development of communities. Nonetheless, art impact studies must be handled with care since they often provide only vague indicators as per their outcomes. This is one of the greatest challenges to be faced when assessing the impacts of arts on social inclusion. According to Sharp et. al. (2005, p. 1006), the impact of public art to the social life of communities and cities is of a symbolic nature rather than a material one. This results in broad criticism regarding the implementation of public art since it is so hard to measure the benefits it is supposed to be having on communities. Moreover, the question arises as to whether the aestheticisation of socially integrative projects by means of art really increases the publicness of urban spaces or whether it in fact represents an aesthetised tool of reaffirming the dominant power's view of the role of contemporary art. One must thus not only be wary of converting art directly into a tool of economic growth, but also of overestimating its function of being socially integrative. Miles (1997) claims that one of the main challenges in advocating art as a social good is its elitist nature: "the exclusivity of taste; the lack of specificity of the public(s) for whom it is intended; and the transcendent aesthetic of modernism is what separates art from life" (p. 16).

Taking Fraser's perspective into account, the question remains – how can public art spaces in which different publics meet and interact in a free and equal manner be realised in practice? Featherstone and Lash claim that artwork, just like academic writing or other intellectual products, may be compared with black boxes located within 'social spaces'. They are deterred from opening up due to their 'ascribed sacred qualities'. This implies that those producing symbolic goods are trying to reach an autonomous level of the cultural landscape and thereby limit their accessibility to cultural products (Featherstone and Lash, 1995, p. 23). Based on this claim, the perspective of art creating a gap between 'high' and 'low' culture as defined by Bourdieu (1984) can be maintained such that art will remain an instrument of 'distinction' rather than a tool of social integration.

To summarise, two contrasting dimensions of integrating the arts into cultural policy and urban planning have evolved as a consequence of globalisation processes and consequent transformations within the economic system, as well as of the growing aestheticisation of urban landscape and daily practices. The question remains as to what extent it is possible to combine these two dimensions and their potential positive effects for economic growth and social inclusion within one particular art centre. Chaney (2002) claims that it has become the responsibility of cultural institutions to act as a mediator between the local and the global level. This claim is based on the development of globalisation not only triggering interurban competition within the cultural sector, but also leading to a decrease of nation-states' influence and an increase of the local-global perspective: "the relationship between citizen and state, at least in the cultural sphere, is shifting from the state encompassing identifiable rights and obligations to acting more as a facilitator of diversity and a mediator between its citizens and global trends and markets" (p. 169). However, referring to Sharp (2005), "the linkages between social inclusion and urban economic competitiveness are disputed" since the objective of social integration "erodes the ability to maintain the competitiveness" (p. 1005) of cultural institutions.

One of the practical consequences of the implementation of this 'entrepreneurial city' model is that spaces where heterogeneous groups of citizens meet have significantly decreased within the urban environment. Moreover, this tendency seems to be indirectly supported by dominant public opinion. Hence, in order to create accessible public spaces in the era of interurban competition and the aestheticisation of urban space, public art centres need to take on the responsibility of fostering communication between individuals, which in turn might lead to the development of new terminologies and perspectives within their publics.

The effectiveness of combining global and local dimensions in this respect is widely disputed; it will therefore be valuable to analyse the structures and experiences of an initiative that purports to combine these two perspectives.

Matadero Madrid within the context of city planning

In order to grasp the rationale of the Matadero's (planned) activities within its urban context, in this section the whole project of Matadero Madrid will be observed against the background of the 'Plan Especial Rio Manzanares' – the Madrid re-urbanisation project in which it is embedded. A barrier between the two riversides of the Rio Manzanares running through the South of Madrid mainly existed due to the legacy of the M30 highway. It constituted a frontier between Matadero Madrid and the riverside as well as between the neighbourhoods on both sides of the river until 2004. However, since then the motorway has been relocated underground. Space was then created on the surface, generating a connection between the neighbourhoods and thus extending the city core (EsporMadrid, 2008).

The 'Plan Especial Rio Manzanares' follows four main objectives:

One aim is to constitute an environmental axis of the city reaching from El Prado until the district of Getafe. The second objective is to organise, develop and urbanise the public space at the riverbanks of Rio Manzanares, especially those areas that were recovered as a consequence of relocating the M30 highway to the underground. This will be realised mainly by the laying out of a park, therewith constituting a physical connection of the two riverbanks. Thirdly, the plan presents a tool to enable a process of adaptation for the citizens of this new public space by the means of providing information and venues of participation, accessibility and the enjoyment of the environment. This objective can clearly be identified as integrating a socially inclusive and participatory responsibility towards the inhabitants of the districts. A further goal is to achieve the citizens' integration by means of alternative leisure into the cultural axis of the city. This includes for example the 'Centre for Contemporary Creation' 'Matadero Madrid'. Finally, the Plan aims at improving the urban integration between the Centre and the Southern and Eastern districts of the city, therewith transforming the river Manzanares moving from its former status of an urban barrier to an ambit of citizen encounter, a connection between neighbourhoods and a conversion of socio-economic components.

Among other establishments and infrastructures, the Matadero constitutes part of the cultural facet of this re-urbanisation project. The 'Plan Especial Rio Manzanares' presents itself as a project that will improve public accommodations and suggest new actions within marginalized neighbourhoods.

The Matadero Madrid within the context of 'Plan Especial Rio Manzanares' is represented as follows:

> With the new use of the Matadero, its cultural infrastructure will open up to the city and therewith provide a growing range of cultural activities to the city's inhabitants. In this sense new opportunities will arise concerning the creation of this new facet of the city by the river (Plan Especial Rio Manzanares, 2008).

Thus, the objectives of the 'Plan Especial Rio Manzanares' explicitly include the social cohesion of districts located on opposing sides of the river and the provision of easy access to cultural activities for the broad public.

Considering the objectives of the plan, a certain social responsibility is officially assigned to the Matadero. However, the authenticity of the urban planners' expressed intention to act for the social benefit of the territory being regenerated needs to be questioned. Based on the trend that city planning nowadays is above all steered by global instead of local impulses, developers and Urban Development Corporations often seem to ignore the needs of the grass-roots level when implementing 'transnational spaces for business and finance' and therewith tend to "redetermine the city amidst its ruin, bypassing the residual (and immigrant) populations" (Sassen, 1996, p. 189). Since global aspirations have become a main focus of urban and cultural policy, projects such as the Matadero Madrid run the danger of being implemented outside the context of its location, disregarding the past and people of the place in which the project is developed. Therewith, the positive impact that the Matadero could have on the social infrastructure of its immediate environment is likely also to be limited. An example of contradictions arising between the history of place and aesthetised urban development is given by Miles (1997), who claims that the dock area of Cardiff "like the London Docklands, [...] is geographically adjacent but separated in ambience and land use from the city, it abolishes the past of the site"(p. 108). Cultural industrial quarter models – or in this case the recycling of the industrial factory – "tend to neglect both the historical precedents and the symbolic importance and value of place and space" (Evans, 2001, p. 41). By creating an abrupt dissonance with the memory of the industrial site and the continuing legacies that surround it, this newly-established art centre might not connect to dwellers' interests due to the fact that it follows objectives of global competition within the art landscape and thereby may lean toward a rather elitist representation of artistic products that offer a narrow scope of accessibility. Miles (1997) describes this conflict as follows:

Art and architecture, then, have coincidental interests which extend from their shared professional ideologies and align them to the needs of developers to create enclaved 'places', but set them against a model of the city in which its dwellers determine or at least influence its values and forms. Whilst at a local level the cultural industries may assist regeneration, though with the danger of gentrification, the corporate development in which art is more often commissioned [...] represents global interests (p.109).

This viewpoint is supported by del Olmo (2004), who in the article 'Poco pan y mucho circo' describes the effects mega-events have on the inhabitants of cities. Although the author refers to events that are organised only on the short-term basis, such as the 'Forum de las Culturas', in Barcelona, or the planning of the 2012 Olympic Games in Madrid, they still follow similar objectives to that of the recuperation of an old industrial site, namely the intention to ameliorate the consequences of deindustrialisation, unemployment and social service cuts. In the case of Madrid – a city with one of the highest rates of vacant dwellings in Europe – a project such as the Matadero suggests a risk that it will not be of any benefit to the inhabitants of 'la Chopera', the neighbourhood in which the Matadero is located. Instead, the aestheticisation of the neighbourhood through the transformation of the old slaughterhouse into an arts centre may lead to its gentrification, implying an increase of housing prices and therewith the eviction of those originally inhabiting the neighbourhood. This in turn suggests a scenario in which "the city becomes uncomfortable for the inhabitant, whose needs are subordinated to the pleasure of the visitor" (del Olmo, 2004, p. 79), the aesthetic eye of the mobile becoming more relevant than the interests of subordinated publics such as the neighbourhood inhabitants. This possible future situation can be compared to the establishment of the 'Forum Universal de lasCulturas', in which the Barcelona government has justified urban transformation along economic interests by creating an aestheticised multicultural event. Del Olmo describes the organisation of these mega-events as "an effective excuse to finally implement urban renewal in the last coastal area of Barcelona that still has a low income population"(Ibid., p. 71). In order to elaborate on the conflict between the goals of the Matadero within the context of cultural and urban policy and the interests of inhabitants living nearby, the following chapters will examine the objectives and actions taken by one of its initiatives, Intermediae, which claims to integrate the local level within the activities it realises.

Good intentions

By examining how the idea of strengthening the collective memory at the local level is carried out, this section will provide further insights into whether Intermediae succeeds in combining the narratives of artistic excellence with a grass-roots integration of the neighbourhood.

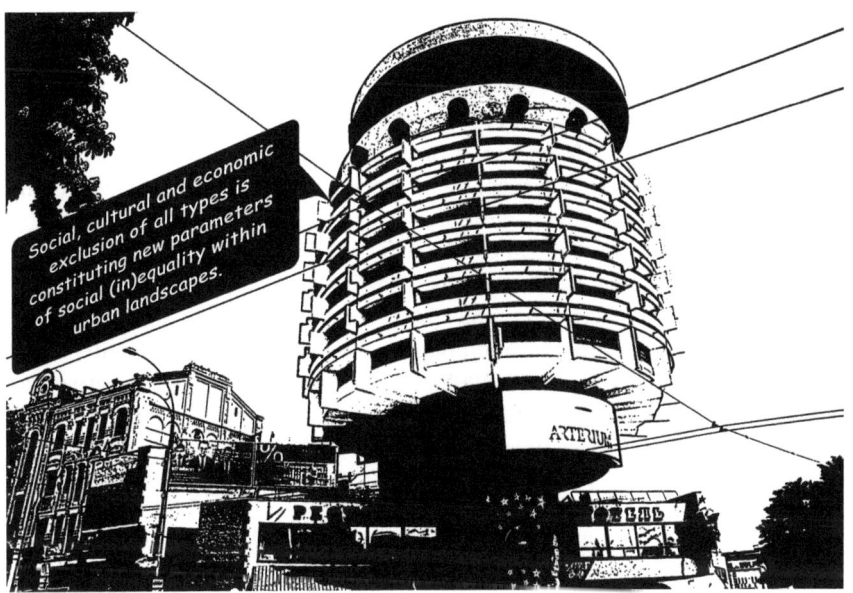

Social, cultural and economic exclusion of all types is constituting new parameters of social (in)equality within urban landscapes.

Objectives of Intermedaie

Chaney (2002) claims that while earlier it was widely accepted that "radical modernism articulated a distinctive and unique challenge to rationalising authority" (p. 163), "now it seems that more generally those working in art world settings [...] have had to consider their social role" (Ibid.). Institutions are nearly constantly obliged to justify why artistic products and events of high culture should be developed and how they relate to a general audience. By implementing the initiative Intermediae into the premises of Matadero Madrid, a little space is reserved for the dimension of the arts that integrates the broader and particularly the local public: one of the main objectives of the "new model of cultural management, promotion and exhibition of contemporary art" (EsporMadrid, 2008) of Intermediae is to connect new creators with the inhabitants of the city. Maria Bella, one of the directors of Intermediae, describes its objective as follows: "We will raise new manners of thinking and to relate oneself between the public and the artist" (Riaño, 2007, p. 36). More specifically, the public is not defined as only composed of those initially attracted to the arts. Instead, Intermediae tries to address all types of visitors, "the young, the specialist, the non-specialist and the local public" (Riaño, 2007, p. 38).

Objectives of 'Procesos de Archivo'

As mentioned earlier, Intermediae finds itself within the complicated situation of being located in a post-industrial factory transformed into a vanguard centre for contemporary creation. Matadero, nearly by definition, runs the danger of neglecting its historical precedents and the interest of the inhabitants living within the district. However, according to Chaney (2002), one strategy for contemporary arts institutions to combine the global and the local level is based on the idea that "global cosmopolitanism has been adapted to the re-energizing of a local identity". Chaney claims that "these cases obviously use famous connections as a peg with which to bridge [...] a distinctive local identity with a global culture" (p.166). Sharp (2005) also points out that by rendering the history of non-recognised communities, one can overcome their marginalised status within mainstream urban histories (p. 1007). According to Griffiths (1999):

> an important part of the experience of exclusion is a weakened or non-existent sense of identity and pride. A key step in integrating excluded populations into the social mainstream, therefore is to assist them to find their voice, to validate their particular histories and traditions, to establish a collective identity, to give expression to their experiences and aspirations, to build self-confidence (pp. 463–464).

'Procesos de Archivo' is a platform of activities initiated by Intermediae along with a project that receives the financial support of the Department of the Arts. 'Procesos de Archivo', was introduced to Intermediae in April 2009. In order to clarify the objectives that Intermediae assigns to 'Procesos de Archivo', an unofficial draft for future interested parties on behalf of the press was consulted. This draft reveals in which manner Intermediae claims to contribute to objectives of both local and international nature. In order to highlight the local function of the project, Intermediae emphasises the platform's nature of being directly connected with the neighbourhood of the Matadero. Moreover, 'Procesos de Archivo' is introduced as a project that aims at integrating the neighbourhood's inhabitants by creating an archive. It is defined as a platform of thought and a collaboration of artistic projects that create a local collective memory of the space 'Matadero' that will be of interest to its various audiences, the objective of which is to elaborate different manners in creating an archive.

With the introduction of this new artistic platform, Intermediae intends to respond to questions such as how one can present and provide access to an institution in regards to the public domain. The action strategies are supposed to be based on the concepts of process and participation. The proposals that are realised within this platform consist, according to Intermediae, of 'living archives' dedicated to collecting and representing a local collective memory as well as a memory of place connected to the Matadero and its surroundings. By claiming that "just as the Matadero accommodates clearly distinctive audiences, the artistic approaches are supposed to represent this multifaceted character", Intermediae reveals that it wants to attract local as well as specialist audiences. Moreover, Intermediae states that "the various points of emphasis that will be worked out in form of different artistic projects are intended to facilitate the transparency of showing common challenges, difficulties and common doubts within the neighbourhood"[43]. This statement refers back to the importance of the platform as an opportunity for local inhabitants to articulate their interests in public. According to Intermediae, the challenges will be worked out with different tools that take on the responsibility of constructing the archives within the context of contemporary culture. By referring to highly prestigious institutions at an international level that are already working with artistic archives – Fudación Antoni Tàpies, The Museo de Arte Contemporaneo de Barcelona, the Museo Nacional Centro de Reina Sofia, the Tate Modern, or the Centre Georges Pompidou – Intermediae emphasizes the significance of the platform on not only the local, but also the transnational level.

To summarise, the marginalised collective history of Arganzuelas' inhabitants is marked by manifold perspectives due to its strong immigrant tradition and therefore presents a multifaceted view of the district. Recapitulating or discovering this collective memory therefore is an adequate alternative to hegemonic narratives or traditional museum archives. In theory, Intermediae has presented a balanced representation that incorporates both its 'local' and 'global' significance. To what extent the interests of a local audience are actually integrated will now be examined by taking a closer look at the nature of the projects actually realised within the 'Procesos de Archivo' platform.

---

43 Information drawn from an unofficial draft of the objectives of Procesos de Archivo, received March 2009

Activities of Procesos de Archivo

One challenge for Intermediae in achieving a horizontal process within the 'Procesos de Archivo' platform is the fact that it may not decide which projects are to be realised within the Matadero. This decision-power is instead assigned to the jury of the Arts Department of the City Council Madrid. Although the concept of creating an alternative archive of the surroundings of the Matadero seems to be an adequate strategy of giving the communities living close to the Matadero their own voice, the nature of each project needs to be taken into account in order to indicate whether they are really representing a participative way of creating a collective memory of the artist, the institution, the creator and the broader public.

There are several activities that have been carried out within the platform of 'Procesos de Archivo'. One of them is 'Museum Futures', which is a project that consists of a video produced in collaboration with the artists Neil Cummings and Marysia Lewandowska and commissioned by the Moderna Museet, Stockholm, in 2008. It presents a virtual interview of a centenary in 2058 with the museum's future executive as interviewee. On its website, Intermediae presents the video as a possible genealogy for contemporary artistic practices and its institutions by imagining the role of the artists, museums, galleries, markets, factories and academics (Procesos de Archivo, 2009, Museum Futures). It is clearly orientated towards the interests of art professionals, be it artists or those working for art institutions. 'Museum Futures' is a representation of hypothetical contemporary artistic practices, leaving little space for participation on the local level and little probability of raising the interest of an average, non-specialist audience. This art project rather seems to create a subject for discussion and provide impulses for action on the topic of the role of future museums by addressing the dominant public, namely art specialists and professionals, in deciding what direction the role of art may take in the future. The channels of other 'alternate publics' may not even be reached on a perceptive level in this project, as the language used in the video is highly academic.

A second activity, 'Future Archives', presented sessions about 'public institutions of the future' in collaboration with Intermediae in June 2009. This artistic collaboration, under the coordination of Manuela Zechner and Valeria Graziano, proposes a workshop in order to create an archive of Intermediae's experience up until now, claiming to include those actors who have been involved so far: collaborators, artists, audiences, associations, collectives and the team of Intermediae. 'Procesosde Archivo' theoretically tackles the challenge of maintaining sustainability within an organisation that is based on a high degree of flexibility and short-term projects. However, a local identity can only evolve if these archives integrate an active, not passive, involvement of the local public. This could be initiated by communicating the projects to the public in a way that encourages them to actively participate and voice their interests in an arts institution. The course description as presented on the website of Intermediae describes 'Future Archive' as a project:

> which approaches a series of solutions to the problem of how to implement distinct futures. It collects conversations based on times and spaces that might possibly evolve and proposals in which one or more persons inhabit versions of the future in a performative way. In this way, the contemporary society is memorised. With each conversation a different future is negotiated with the help of a discourse method that borrows methodologies of militant investigation and radical pedagogy, as well as with the help of interviews and dialogue. By intending to offer spaces in order to carefully develop vocabularies and gestures that will be able to indicate potential manners of thinking, acting and existing, Future Archive involves the articulation of hopes and wishes for future modes of ccoexistence, negotiating the space between a recorded present and a potential future as well as confronting the challenges arising out of current proposals and imaginaries (Procesos de Archivo, 2009).

Such a project description leaves little chance for the average local visitor to be lured in by the course, since its content is explained in a highly conceptual and academic manner and rather seems to follow the dictate of establishing an intellectual, art-specialist image as corporate identity for the 'Matadero Madrid' as a prestigious art space instead of actually aiming at catalysing the active participation of the (local) public within the workshops. The problem that arises in working with such a corporate image is that there probably will be no initial self-interest for the local public to participate in the workshop. Connections to schools or other audiences will have to be established, leaving the 'bottom-up' channel to the local, 'alternative publics' untouched.

A third activity, 'Corte Arganzuela' is presented as a 'documentary theatre project' of the Company PUCTUM that works with transformations of the districts and its neighbourhoods by telling the life stories of its inhabitants. "[PUCTUM] reconsiders the dynamics of a scenic laboratory [...] and incorporates new neighbours by a new call of applications directed to the community of Arganzuela" (Procesos de Archivo, 2009). PUCTUM actually chose a suitable way in communicating their project to the neighbourhood of the Matadero. By spreading handouts in bars where the older, local neighbourhood likes to gather, talking directly to people and presenting the idea of PUCTUM and even involving children playing on the premises of Matadero, the artist Cecilia Pradal succeeded in directly communicating the project to an audience that usually did not visit the 'Matadero Madrid'. After a casting in which the applicants showed their talents, a group was selected to perform a stage play about their life stories and the story of the Matadero itself. Brainstormings were held with the locals, during which the experiences and problems of the local public were recorded. Although only six people were staging the play in 2008, more than 50 people participated in sharing their ideas, memories and experiences that were later on integrated into the play. Thus, by incorporating, not dominating the language and cultural codes of the locals, a channel for 'alternative publics' to be actively involved within the art space was made possible:

> the inhabitants of the neighbourhood themselves, are working here in Intermediae, like the guys who dropped by last week in order to do the casting, or those that will come for the auditions, which will be right here, and this obviously because the space owns the logic of being in an open condition and it necessarily opens up the process, and in some moments we will open it up and in other moments people will come to spy or simply watch. The neighbours that participate are selected by us, but they don't play against us, on the contrary we are discussing and developing the piece together (Interview with artist Cecilia Pradal, April 17th, 2009).

According to Sharp (2005), state-sponsored projects (such as those of Intermediae) often include a community element and the artists suggest "that their work [within the framework of projects subsidised by the government] establishes a conversation between the spaces and the people who inhabit them" (p. 1016). This is the case with the 'Corte Arganzuela 09' documentary theatre project, whose objective is "with the help of their own [the dwellers'] life stories and their own historical cuts of life etc. to present one possible way of collecting a neighbourhood's memory, which contributes to this 'collective memory' or 'collective initiative'" (Interview with artist Cecilia Pradal, April 17th, 2009). According to Lewitzky (2005, p.52), it is the responsibility of cultural producers and artists to represent the interests of different publics by visualising and realising their ideas, to create access to dominant public discourse and to create participatory possibilities to create alternative urban space. This seems to have been the intention of the artist when considering her performative approach in giving voice to the neighbourhood.

Finally, Intermediae organises 'working sessions' in order to reflect about limits, challenges and responsibilities of a collectively created archive such as the one of Intermediae. Various guests (specialists in the field) were invited to discuss the concept of (artistic) archives with Intermediae members (Procesos de Archivo, 2009). The 'working sessions' disregard the initiation of an actual horizontal process and the attempt to create a platform of inter-public relation.

In order to give a voice to the local inhabitants, it would be appropriate to invite not only specialists in the field of creating artistic archives, but also dwellers or 'ambassadors' of the neighbourhood itself in order to avoid letting the concept of collectivity be dominated by the ideas of an artistic and intellectual elite.

To summarise, Intermediae finds itself within the complex situation of being obliged to assist the realisation of artistic projects that are not chosen by the initiative itself and therefore might not be realisable within the framework of a participative process. Still, those projects that are a result of Intermediae's project planning often seem to equally disregard a horizontal inclusion of the neighbourhood. Although the local level is included in the projects as a subject matter, many of the activities do not go beyond a theoretical discussion of local developments and realities. Moreover, activities initiated by Intermediae do not catalyse an active participation or bottom-up input of the general public. An exception is the theatre PUCTUM, where each actor that voluntarily wanted to participate in the play could tell his or her life story in the way he or she was able to do it best. Unfortunately, the information on the platform's website is additionally often concealed by highly complex artistic and theoretical narratives that complicate the communication between Intermediae and the general public. According to Miles (1997, Foreword), there are two possible functions for art within the context of public realm: "as decoration within a revisioned field of urban design in which the needs of users are central, and as a social process of criticism and engagement, defining the public realm not as public sites but as complex fields of public interests." Considering the above made statements, Intermediae claims to fulfil the second role Miles assigns to the arts in relation with urbanisation. Nonetheless, aside from the project of PUCTUM, Intermediae seems to be merely adding to a socially integrative corporate identity of 'Matadero Madrid' rather than opening 'bottom-up' venues for 'subaltern counter-publics'.

Picture 1: Project planned until the end of 2010: Establishment of two pedestrian overpass and leisure park. On the right riverside, old buildings of Matadero.

Conclusion

As could be observed when examining the objectives of the 'Plan Especial Rio Manzanares' and 'Project Matadero Madrid', the urban planning project clearly purports to include the aim of strengthening social infrastructures in and between the districts surrounding the old industrial site. A creation of an art centre was a part of the idea aimed at social inclusion. It would not only act as a connection between local interests and global narratives of artistic excellence, but also contribute to the creation of a both symbolic and real bridge between the two riversides and thereby increase the social integration of those districts that, until now, hardly connected to the cultural richness of Madrid's city centre. According to an interview with Pablo Berastegui, the coordinator of Matadero Madrid, the most urgent task for the organisation of the Matadero was to turn it to a project that is well-known to its audience:

> [One of the most important tasks is to] give it more visibility, it is a wonderful project but it is not very well-known yet [...] It is a raw diamond. I hope to achieve the city's inhabitants feeling that it is their own, that they get used to coming here a lot (Vilar, 2008).

However, so far it mostly seems to provide venues and access to the members of the highly-aesthetised and intellectually profound world of contemporary artistic creation and creators, while the majority of the city's public remains outside the brick walls of the slaughterhouse complex. With a lack of channels to reach and activate not only the 'dominant public' but also 'subaltern counter-publics', the Matadero will not be able to escape the exclusiveness that might be closely connected to a global, but not a local audience, particularly in a district marked by a strong workers' and immigrant tradition. In this case doubts arise concerning the publicness of public art space for the dwellers living in the district. A public art space will only be able to create new synergies between different publics if a horizontal venue of discourse will be created that can alternate and actively influence the projects carried out within the framework of Intermediae.

The initiative clearly pretends to apply a social dimension to the arts. However, an elitist, artistic narrative is currently dominating the projects. Intermediae aims at including the broader public but at the same time runs the danger of instrumentalising the concept of the 'local' in order to create aesthetised sceneries of fake horizontality. Within the framework of the Matadero, as part of a re-urbanisation project, this raises the suspicion that Intermediae was established just due to "the government's need to portray a peaceful and pleasant image of the city" (del Olmo, 2004, p. 72). Although the implementation of Intermediae's working philosophy is exacerbated by the project's complex administrative structures and carries the difficult task to combine various dimensions of and interests in art, some projects such as 'Corte Arganzuela '09' have achieved first successes in leaving spaces for a 'bottom-up' participation in artistic projects. However, the initiative and the artists themselves have not yet delved into the many possibilities for creating a more participative platform.

It is thus crucial for various actors involved in developing 'contemporary creation' within the context of urban development – for the Intermediae managers, the artists, the Department for the Arts of the City Council, as well as for the city publics themselves – to undertake the decisions and activities leading to closing the gap between newly established cultural sites and the life of the city dwellers. This can only be done, however, if those involved in the artistic process create possibilities for those at the grassroots-level to actively make decisions in creating and recycling urban space.

References

BLAKE STEVENSON LTD (2000) *The Role of the Arts in Regeneration.* Edinburgh, Scottish Executive Central Research Unit.
BOURDIEU, E. (1984) *Distinction: A Social Critique of the Judgment of Taste.* London: Routledge.
CASTELLS, M. (2000) The rise of the network society: The information age: economy, society and culture, Volume I. 2nd ed. Oxford: Blackwell.
CHANEY, D. (2002) Cosmopolitan Art and Cultural Citizenship. In: *Theory, Culture & Society, Vol. 19*(1-2), pp. 157–174.
COMMUNITY DEVELOPMENT FOUNDATION. (2001) Social Inclusion and Community Development Practice. Available from: www.cdf.org.uk/html.socinc.html.
ESPORMADRID. (2008) Pasarelas Matadero I y II. Available from: http://www.espormadrid.es/2008/06/pasarelas-matadero-i-y-ii-s-2-y-s3.html.
EVANS, G. (2001) Cultural industry quarters: from pre-industrial to post-industrial production. In: BELL, D. & JAYNE, M. (eds.) *City of Quarters: Urban Villages in the Contemporary City.* Aldershot: Ashgate, pp. 71–92.

FEATHERSTONE, M. & LASH, S. (1995) Globalization, Modernity and the Spatialization of Social Theory: an Introduction. In: Featherstone, M., Lash, S., Robertson, R. (1995) *Global Modernities*. 1st ed. London: Sage.
FRASER, N. (1992) Rethinking the Public Sphere: A Contribution to the Critique of Actually Existing Democracy. In: CALHOUN, C. (ed.) *Habermas and the public sphere*. Cambridge, MA/London: MIT Press, pp. 109–134.
FRENDE VEGA, M.A. & PENA SANCHEZ, A.R. (2010) *Las Empresas Culturales y su impacto en el crecimientoecononómicoregional*. Conference paper presented at: International Meeting on Regional Science. The Future of Cohesion Policy, 7th Workshop, Badajoz-Elvas, November, 2010. Available from: http://www.reunionesdeestudiosregionales.org/elvasBadajoz2010/htdocs/pdf/p235.pdf.
GRIFFITHS, R. (1999) Artist organisations and the recycling of urban space. In: NYSTRÖM, C. (ed.) *City and Culture: Cultural processes and Urban Sustainability*. Karlskrona: Swedish Urban Environment Council, pp. 460–475.
HABERMAS, J. (1989) [1962] *The Structural Transformation of the Public Sphere: An Inquiry into a category of Bourgeois Society*. Polity: Cambridge.
LEONTIDOU, L. (2003) *Spaces of Risk, Spaces of Citizenship and Limits of the 'Urban' in European cities*. Presented at: (In)visible Cities. Spaces of Hope, Spaces of Citizenship, Centre of Contemporary Culture of Barcelona July, 2003. Available from: http://www.publicspace.org/es/texto-biblioteca/eng/a021-spaces-of-risk-spaces-of-citizenship-and-limits-of-the-urban-in-european-cities.
LEWITZKY, U. (2005) *Kunst für alle? Kunst im öffentlichen Raum zwischen Partizipation, Intervention und Neuer Urbanität*. Bielefeld: Transcript Verlag.
MAYER, M. (1994) Post-Fordist City Politics. In: Ash Amin (ed.) *Post-Fordism. A Reader*. Oxford: Basil Blackwell, pp. 316–337
MILES, M. (2005) Interruptions: Testing the Rhetoric of Culturally Led Urban Development. In *Urban Studies, Vol. 42* (5/6), pp. 889-911.
MILES, M. (1997) *Art, Space and the City: Public Art and Urban Futures*. London: Routledge.
OLMO DEL, C. (2004) Poco pan y muchocirco: el papel de los 'macroeventos' en la ciudadcapitalista. In: *Archipiélago: Cuadernos de critica de la cultura, Vol. 62*, pp. 69–80.

PLAN ESPECIAL RÍO MANZANARES (2008) Distritos de Moncloa-Aravaca, Centro, Arganzuela, Latina, Carabanchel y Usera. ANM 2008\27. BO. Comunidad de Madrid 31/07/2008 num. 181. Pp. 84-102. Available from: http://www.madrid.es/boletines-vap/generacionPDF/ANM2008_ 27.pdf?idNormativa=abe1b29e6a7bc110VgnVCM2000000c205a0aRCR D&nombreFichero=ANM2008_27&cacheKey=18
PROCESOS de ARCHIVO. (2009) *Corte Arganzuela* '09. Available from: http://intermediae.es/project/intermediae/blog/procesos_de_archivo_cia_puctum _corte_arganzuela_09 [Accessed 02/06/09].
PUSEY, M. (1987) *Jürgen Habermas: key sociologists*. London: Routledge.
RIANO, P. (2007) Armar Lo Imposible. Calle 20: La Revista de la NuevaCultura, 13, pp. 36–39.
SANCHEZ, S. (2008) *Sobre las posibilidades de construccion de unainstitucionpublica, pericepias e intentos de gestion cultural*. Presented at: Practicas mas alla y a pesar de la sospecha, Sao Paolo, 2008.
SASSEN, S. (1996) Analytic Borderlands: Race, gender, and representation in the new city. In: KING A.D. (ed.). *Re-presenting the City: Ethnicity, Capital and Culture in the* 21st Century, New York: New York University Press, 1996: 183–202.
SCOTT, A. J. (2006) *Creative Cities: Conceptual Issues and Policy Questions*. Journal of Urban Affairs, 28: 1–17. Available from: http://scholar.google.com/scholar?q=Creative+Public+Policy+space&hl= de&lr=&start=10&sa=N.
SHARP, J., POLLOCK, V. & PADDISON,R. (2005) Just Art for a Just City: Public Art and Social Inclusion in Urban Regeneration. In: *Urban Studies, 42* (5-6), pp. 1001–1023.
VILAR, C. (2008) Yonosoyun 'cultureta'. Available from: http://www.elmundo.es/papel/2008/11/04/madrid/2535712.html [Accessed 26/05/2009].
WILLIAMS, D. (1997) Creating Social Capital. Adelaide: Community Arts Network of South Australia.
ZUKIN, S. (1995) *The Cultures of Cities*. Oxford: Blackwell.
ZUKIN, S. (1998) Politics and aesthetics of public space: The 'American' model. In: *Ciutat real, ciutat ideal. Significatifuncioal'espaiurba modern.* Barcelona: Centre de CulturaContemporania de Barcelona, Urbanitats, 7. Available from: http://www.publicspace.org/en/text-library/eng/a013-politics-and-aesthetics-of-public-space-the-american-model [Accessed 02/06/09].

CHAPTER 4

Tracing Art in Urban Public Space: the Resistive Aesthetics of Cultural Actors in Post-Communist Romania

Laura Panait

> *Don't Marx me* (phrase used by the
> Romanian artist Dan Perjovschi)

I intend to explore in this chapter how notions of public space can be operated and investigated in the Eastern European realm of Romania. Toward this aim, the "radiography" indicators I will use as testing elements of the urban tissue are cultural and artistic actors.

Nowadays, speaking of public space means being particularly aware of the debates on the issue of public realm and "publicness", and having this context in mind as a major interpretative framework. A significant feature of this context is the multidisciplinary approach. In order to hold the viewpoints of architects, geographers, urbanists, designers and also artists, it is crucial to stay open-minded while penetrating this complex subject of urban public life and space. The discipline of anthropology works nicely in trying to identify certain tendencies in this social space, but seeing this constellation in a more sophisticated way requires constant collaboration with other actors who observe or act on this side of the city. Furthermore, even when focusing on certain communities, one must keep constant contact with other disciplines in order to really understand the causes, effects and symptoms of processes that are taking place within a city.

Due to the diversity of analyzed themes related to both public space and the artistic sector in the city, I emphasize here the relevance of an interdisciplinary approach. Methodologically, my notes are based on the analysis of visual and theoretical documentation and on data collected in the course of participant observation. Additionally, I will dwell on ethical issues and, in particular, on the method used in the interviews and on the other inter-relational aspects of anthropological methods.

One of my main motivations to start working on the relationship between cultural actors within public space was the discovery of certain data published by the Center for Studies and Consulting in the Cultural Domain (CCCDC, 2009) in their study, entitled "Culture in public space. The analysis of the artistic events taken place in Bucharest", they state that the "number of events organized by NGOs in public space nearly grew by three times", from 9 events up to 25 at the end of the studied period. Therefore, "a sustained growth characterizes the presence of the non-governmental organizations as events' organizers within the public space" (p. 6). I consider their study as being a strong starting point for my further research, confirming statistically that more and more cultural actors became increasingly active in this space, creating new opportunities for analysis but also for action. Together with this research, a series of debates on public space were organized a few times in the last few years, including the role of art as activator within this context.

I also had the opportunity to participate in numerous projects involving artists wishing to provoke the public within the most common spaces of the cites of Cluj and Bucharest. I realized that throughout all these projects I would have to place myself both as insider and outsider. However, I decided that taking part in the events as a participant in artistic interventions in public space would give me a better chance of being accepted by the local art community than as an external observer. I found the participant observation, working with and analyzing these interventions, to be a valuable achievement.

The interviewing was one of the strongest elements of my research. In more than 30 interviews I tried to find out exactly who was my community to examine and who would be the key persons to whom I should speak. By interviewing a variety of participants in the art scene – from people in the neighborhood, artists, and researchers to the audience itself and the city authorities – I tried to cover a wide range of possible answers concerning my theme and gain portraits of the cultural actors involved.

What are the public spaces for?

When I began analyzing 'public space', my idea of it was a city space desiring to be rediscovered by city inhabitants through daily activities, a space constituted by the relationship between different actors shaping the life of the city, including artists and artistic expression as a particular and significant aspect of public space.

After this motivational moment on the topics of public space and art, I started to see the Romanian situation through the CCCDC (2009), conducted between 2003–2008, whereby the concept of public space was applied both "to the analysis of a geographical space throughout the evaluation of the way of access to it and to the analysis of what we call the space of confrontation of ideas in society" (p. 8).

Urban design often pursues the clear-cut instrumental goal of comfort, practicality and order. But the scope of everyday life in urban spaces is never completely subordinate to the achievement of predefined, rational objectives. People can be sometimes capricious and unpredictable. Urban spaces and their activities constantly generate disorder, spontaneity, risk and change. Urban public spaces offer a richness of experiences and possibilities for action (Stevens, 2007, p. 1).

Public space is limited and unlimited at the same time. Its limits are given by the private sphere and its non-limits exist through the continuity offered by the succession of public spaces. The public space is never insular, but only intensified. From this perspective, the public space can be compared to nature (Mihali, 2005, p. 108).

Urban experience and social needs are more than mere conceptual abstractions; they can be understood by looking at everyday life on the streets, at its specific and diverse qualities and at the meanings it may hold for those who live it, and also by looking at the complex tensions which arise between different needs, meanings and users in spaces. Henri Lefebvre, Walter Benjamin and also the Situationists all identified social practices of people as a key to understanding the dynamic tensions that shape everyday life in public. Lefebvre distinguished how cultural life and social life struggle to find their realization through urban spatial conditions (Mihali, 2005, p. 109).

Asked about the idea of post-contemporary public space, Boris Buden (2003), a Croatian theorist who made significant input to the debate on the issue of public space, finds the source of these thoughts within "the Kantian notion of using reason publicly, the courage to articulate one's own ideas openly and publicly. Of course, these ideas have been reconfigured according to the social and political transformations in the last decades" (pp. 3–5).

From this general overview, looking closely to what is happening today in Romania, Ciprian Mihali (2005) compares quite drastically the relationships in public space with a cheap theater play, with second hand actors in which the "city of today resembles less a scene and more a screen, transparent, isolating, cold and lacking deepness, on which is shown as in a shopping window with TV sets, repeating some episodes from an ordinary soap opera" (p. 99).

In this context the philosopher is assuming that:

> the value of public space is getting smaller within the authenticity and personal quality of language, the freedom so popular with the beginner modernists becomes desperate liberation in the others, in its pure de-responsabilization of action lacking consequences (pp. 99–100).

Mihali underlines here what his generation represents[44], including a group of theorists and artists who are critics of individualism and over-consumption in the transitional period. They see this obsessive need being transferred into everyday practices of public space, dominated now by the new economic connotations.

"Post-communism" as the Romanian condition and the notion of "public space"

Looking into the different historical layers of Eastern European realities, we can assume that public space had different dimensions here. To provide a foundational framework, I will present the arguments of several contemporary Eastern European theorists who explored what post-communism and public space are and how we can deal with the comparison of these national contexts today.

A way of starting the discussion is by introducing the Serbian contemporary philosopher Miško Šuvaković (2007) and his work "Apocalyptic Spirits: Art in the Post-socialist Era", offering the typology of actual macro-cultures. He defines post-socialism as "postmodern, post-historical and post-ideological culture" which "is determined by crossing over, lateness, regression and loss. [...] Post-socialist countries are entropic cultures. Cynically speaking, they are entropic cultures, because they are based upon premodern or modern production and postmodern consumption" (pp.134–141).

But when speaking about effects, the Romanian theorist Ciprian Mihali (2005) underlined that:

> communism did not create public space (or private either but not only because of this), but the after era, militarily and economically aligned to a new order, seems less willing to take care of something public, even if it were to be in the name of human rights, liberty and other of modernity's export goods (p. 100).

In this way, individual development for many Romanians is not connected to common space and even less to public space. In the 1990s, Romanians who had been lacking private life for decades, started enjoying all the benefits of it and the freedom they did not have before the Revolution. However, the public realm, public life, and space, were still ignored for a while. The lack of balance

---

44 The group from Cluj gathered under the Idea Publishing House, the only left-wing publishing voice in Romania, which used to have an exhibition space called Protokol; they also have an important critical contemporary art magazine: http://www.idea.ro/editura/.

between investing in public space and one's own privacy became increasingly visible.

On the other hand, Boris Buden (2003) stressed that "even in communism there was a public sphere, (...) ideologically closed one, subjected to different sorts of direct political control". He speaks also about "the so-called space of cultural hybridity, the space of mixture, a new post-essential, post-national culture" (pp. 3–4).

Speaking in particular of Romanian public space, I would like to refer to an influential Romanian architect from Cluj, Şerban Ţigănaş (2005) who identified certain 'mutations' applicable to public space and represented in the Romanian landscape in the last decade. Here are several examples from Ţigănaş, in regards to Cluj. The first mutation is "Change its name!", referring to the dispossessing of memory. The second is the "Occupy it with symbols!" mutation, in which the administrative and political powers take into possession the central hubs of the space and fill it with "monuments, statues, commemorative stones (...)" which might change the character of the space imposing also other forms of censorship. The third one is "Change its limits!" by redefining the owners of the public space. The fourth one is "Create the public space!", as a "generator of the coagulations of private development or of the partnership between public and private, the mutation being a beneficial contemporary instrument that needs a capable project management". And the last mutation is "The big project!", as in "the big approach", which if not well implemented creates blockages in public space. A good example that Ţigănaş gave is the Greek-Catholic church that for over 18 years has received a space in the city center, is still not finished and simply physically blocks the space. The architect's answer to questions about the date of finalization for this project is simply cynical: "It will be finished when mixed with a supermarket!" (p. 101).

Occurring over several stages, the 'emancipation' of the post-communist society was accomplished at different levels. However, when they intersect they produce hybrid cultural manifestations. One of these is found in contemporary art. When observing the configuration of public space, with a focus on what the transition after the 1989 Revolution meant for it, we have to look upon the role of contemporary art in trying to shape post-communist reality. It attempted to perform as a catalyst of rebirth for civil society, which implied particular activities in the public sphere and some changes in urban public space. Contemporary art inserts itself into public space in a moment when it was much in need, when the media was not reflecting many significant aspects of the life of urban public spaces and when the conscience for this space was still quite unstable.

Art in/through public space

As I have shown, public space in a post-communist and in particular Romanian context still shows the symptoms of 'weakness', but also of recovery after the long process of communism. The role of artists in this process of recovery of public space in former socialist cities seems to be underestimated and under-investigated thus far, at least in Romania. In the following sections, with examples used from art projects performed in Bucharest and Cluj, Romania, I will try to shed some light onto artistic activities in public space and on their impact.

When mapping art in public spaces we first need to make a significant distinction. Artists working for the sake of the public interest address a wide range of human concerns, which could mean that art as a social practice occupies non-gallery sites, i.e. public spaces. On the other hand, art outside the gallery may still be performed more in the interests of the artist or curator than that of public. In other words, the notions of 'art' and 'public' do not fit each other easily or naturally. A definition of 'public art' is fraught with the contradiction that while modernist art has occupied the hermetic space of the white-walled gallery, art forms more closely linked to areas of everyday life, such as "community arts" or "outside art", have been marginalized by the art establishment as lacking "aesthetic quality" (Miles, 1997, pp. 52–53).

Another aspect of the present debates on art regarding public space is encountered in the work of Arlene Raven, entitled "Art in the Public Interest", where she begins asserting that "public art is not a hero on a horse anymore", and that "art in the public interest deserves instead more attention as it extends the possibilities of public art to include a critique of the relations of art to the public domain" (Miles, 1997, pp. 100–101).

Furthermore, with regard to the idea of street art, Malcolm Miles (1997) reaffirms another concept, which has to be taken into consideration when talking about tendencies in public spaces – the art intervention. He states:

> art as intervention in the public realm is a form of continuing social criticism which resists the institutionalization of conventional public sculpture. Its roots range from the social sculpture of Joseph Beuys to 1960s happenings and the influences of Marxism, feminism and ecology; the strategy involves a redefinition of art as a critical realism which does not record urban experiences but seeks to change them according to ideas of social justice and community, for many artists beginning with a personal transformation and relinquishing of the notion of the artist as hero, bohemian or victim (p.13).

Accordingly, the artistic intervention concept applied to the Romanian context was revealed in the study conducted by CCCDC (2009). This research suggests two ways of distinguishing the art events: first, from the time perspective and second, from the physicality and the space itself point of view, i.e. temporally and spatially. Temporality "is a strange idea for architecture and urbanism", disciplines used to build objects and even environments especially from a Romanian point of view, which is not highly aware of ephemeral interventions as another face of creating urban situations. Therefore, artistic intervention provides another perspective of the artistic act in the city and in time reflexivity. In addition, the mentioned research points out that "the temporality is seen in the Western practices as a quality of the space more than a definition of the duration of the intervention within the certain urban space" (CCCDC, 2009, p. 10).

Temporality and artistic interventions are two terms that go together quite fine. The so-called "injection into the public space" is an act meant not to last long in time but to transmit a message that will start some reactions/effects among people and their consciousness. The temporary usage of space is a statement against the permanent architectural-building model, which is imposed on the city-body monumentality and sometime even rigid spaces in the everyday lives of its inhabitants. Regarding this, the temporary artistic interventionists come to tinker with a certain perception of time, space and message. In this way, although the Romanian public space continues to own its own structure and objects, charged as they are with post-communist official discourse, the artistic intervention enters into the scene, from time to time, to change the memory, the usage and the future of this space previously only officially accepted and programmed for certain utilities.

As for spatiality, the authors of the report stated:

> despite the apparent conflict that appears within the temporary usages of the urban space and the represented interests of the urban planners, the relationship between power and artists that interfere in urban space is, in the majority of the Western bigger cities, one of tacit collaboration based on tolerance. This relationship can be compared with the one between the strategies and tactics within a war (CCCDC, 2009, p. 20).

When we compare these assumptions with the ones articulated by De Certeau (1984), we can clearly see the similarities in regards to art related to power and tactics related to strategies. As De Certeau described, "the tactic is an art of the weak. Power is bound by its very visibility. In contrast, trickery is possible for the weak, and often it is his only possibility, as a 'last resort': The weaker the forces at the disposition of the strategist, the more the strategist will be able to use deception." Therefore, "a tactic is determined by the absence of power just as a strategy is organized by the postulation of power" (pp. 37–38).

We will now move on to explore the Romanian examples in particular, keeping in mind the theories and studies presented above. The following Bucharest and Cluj cases are all subjects of artistic interventions to which we can apply indicators such as time, space and tactical approach to the city.

The Cases: Bucharest and Cluj voices

As I have observed in Romania, artists who plan to express social or political ideas and have a strong voice in their domain, prefer to speak out loudly, often doing so in public space and not only between an institution's walls. Such artists ought to be considered 'ice breakers' especially after the year 2000, due to the fact that very few Romanian artists are initiating artistic projects or social processes in Romanian public spaces. Still showing symptoms of the communist censorship that erased much free public expression, the Romanian artist who dares to face the public space must choose 'hot' issues in order to attract public attention. As Quentin Stevens (2007) argues in "The Ludic City", artists are one of the types of players in a city who perform quite a lot and develop skills of playing in and with the city. But after 40 years of restrictions in expressing creativity in general, and particularly in an 'outdoor' manner, the playful disposition and 'social mood' has to be restored. Therefore, Romanian society has to learn again how to use the city's spaces as an immense playground of sorts, to learn the ways of expressing its freedom in public spaces, and overcoming their previous 'watch-dog' attribute.

In the last five years I have observed various different events organized and run by artists and cultural organizations in the public spaces of Romanian cities, such as Bucharest, Sibiu and Cluj. I will limit myself in presenting only two of these cases. These two projects represent two tendencies of art in public space in two cities: "Public Art Bucharest 2007" and "Samples" in Cluj in 2008.

"Public Art Bucharest 2007"

"Spațiul Public București/ Public Art Bucharest 2007", as the curator Marius Babias (2008) was presenting it, "consisted of interventions, debates and actions throughout the year 2007". The project, as a partnership between the public actors and the NGOs, created a platform for interdisciplinary discussions and debates, exploring how public art encourages a critical engagement with the power structures that are dominant in the public sphere. 'Public art' in Bucharest is not merely equivalent here to the monuments, but extends its limits far beyond the conventional understanding of arts; it became a measuring unit for the cultural and democratic state of society. "The artist's projects confronted the public with a series of contemporary themes relevant both from an international perspective and for a context in which the exercise of democracy has not yet been fully incorporated. The streets, squares and markets of the city, public and private institutions, public transportation and mass media channels constituted settings for artistic interventions" (p. 14). This task was solved at the first stage of the project primarily by Romanian artists with an international background, namely Mircea Cantor, Daniel Knorr and Dan Perjovschi.

The particularity of Bucharest is an utopia of a failed socialism that was aimed at producing a human being of new sort and it can be read in Bucharest's architecture and city planning quite clearly. At the same time, currently the wild flow of capitalism is flooding the city; public areas are being privatized, legislation goes unheeded and city planning regulations are simply ignored while the city is rebuilt according to the needs of capital. Bucharest has adopted a new utopia – the one of money (Babias, 2008, pp. 14–15).

Therefore, Bucharest is one of the fastest developing cities in Europe, a place where post-communism and globalization have caused significant frictions in architecture, city planning and social policy. Public space is defined here mainly through social interaction of the inhabitants of the city, through the way they perceive and react to the new tensions. Cities like Bucharest are facing the challenges of unsolved issues of the recent past backdropped by the speed of unlimited capitalist expansion (Babias, 2008, p. 8).

One of the ways to accomplish this task is by bringing these issues into the public debate and interventionist art performed in the public space.

As Babias (2008) again stated:

> Bucharest needs to be recognized as a European cultural metropolis and as such it needs structures that can stimulate exchange of ideas in tune with contemporary international developments. (...) Thus, the events of the Bucharest festival of 2007 had three objectives: to bring Bucharest in line with developments in contemporary art and to create awareness about the importance of the public space, the creation of a self-sustained initiative regarding public art in the mid-term and a continuous realization of public art projects in the long-term (p. 17).

Witnessing the multiple art interventions that were part of "Public Art Bucharest 2007", I discovered that some of them had quite well-structured concepts, such as those based around the main themes of Romanian post-communism. For instance, Dan Perjovschi's "human statue" (see fig. a) simulated the moment of the "mineriadas"[45] from the 90's with a few art students, while Daniel Knorr

---

45 The term "mineriada" defines two social movements from 1990, 1991 and 1999 of Romanian miners, coming from Valea Jiului to protest in Bucharest. These episodes

redesigned several trams with stickers that pointed out the main institutions of the state by lending them their own advertising slogans, such as for the Orthodox church, the Army, the Police etc. Using issues like the miners' riots, the importance of the Orthodox church and urban rituals that function as hybrids between the city and rural practices (such as the lamb sacrifices for Easter performed in the same project by the artist Mircea Cantor and entitled "Silence of the Lambs"), the project aimed to illustrate the complexity of the Romanian context, fragmented in so many directions like the city of Bucharest itself, a puzzle-reality, sometimes hard to understand as a whole.

Overall, the entire Bucharest art project was "illustrative for expanding the geographical area" of the city life: the events organized by the public and independent actors "have covered the biggest area from inside the city, exceeding many times the central axis going to the peripheral areas of the city" (pp.13-14), as the CCCDC (2009) study concluded. Here we find a contradiction which contemporary public art rarely solves. On the one hand, the revitalization of the neighborhoods involved and the stimulation of the city's vitality were among the tasks of this type of intervention in public space. However, quite often public art projects are not highly visibile to the local communities to which they pretend to be addressed. It happened to the majority of the art projects of the "Public Art Bucharest 2007" as well. Except for those in the artistic community, which was quite interested in the proposed projects, individuals from the areas on focus rarely reacted to or were even informed about the presence of art in their neighborhoods. And yet, even if the projects did not create a special relationship with the public in those neighborhoods in which they were performed, it was still an important step for the inner artistic community toward raising its awareness of the possibilities of exploring the public spaces in the city.

"Samples", Cluj, 2008

Multimedia project "Mostre/Samples" (see fig. b) from 2008 in Cluj had its own way of dealing with different tendencies regarding how artists see the city and how they want to interact with it. Furthermore, this project is not about the

---

were set against the government but also against the intellectuals from the city, resulting in the beating of "people wearing glasses" on the streets and the devastation of a few parts of the city center. These "mineriades" are considered to be the cause by bad negotiations between the government and the miners, as a new consequence of the unemployment policies initiated in that time in Romania. They are also informally considered to be, after the 1989 Revolution, the biggest and most brutal events in Romanian post-communist history.

capital of the country, but rather a city with a different historical and cultural makeup while sharing the same post-communist background. In this respect, the city manifests certain similar symptoms in terms of structure, organization and appearance of socialist architecture, especially within neighborhoods, but these symptoms are less visible than in Bucharest. Because the former totalitarian regime did not reshape the city center like the one in the capital, these appearances created a different context for issues linked to center and periphery.

The city still is bound to ethnic discourse, due to the presence of Hungarians composing twenty percent of the total city population and also other smaller percentages of Roma, Jewish and Saxonian populations. Issues linked to the conservation of national symbols come from the Romanian ethnic majority and are expressed especially toward the Hungarians. This manner of resistance to 'the other' made Cluj's public space at times a battleground between majority and minority, using the (historical) monuments as tools of power reaffirmation and legitimisation. At the same time, it transformed the city center, more compact than Bucharest's, into a place where control and national identity are often applied and reaffirmed.

Considered to be conservative but, according to others, also creative place[46], Cluj has a different rhythm that Bucharest in approaching cultural events, at a smaller scale and with different concepts. In this context, few artists after 2000 seemed to have strong attitudes towards this national discourse. And here I would name just one big project, "Mind Bomb"[47], which was oriented against political-electoral-centered discourse. Apart from this, other actors, such as AltArt, focused more on the neighborhoods themselves, as a reaction also to the density of official manifestations within the city center. Based on smaller artistic communities, art is "served" here in small portions and only to the more initiated artistic circles.

In this case, AltArt foundation, who initiated "Samples" decided to explore public space more deeply and to focus on the social issues. They chose to stage

---

46 Cluj is already considered by many cultural operators as being a cultural and creative city. But on the other hand, some of my interviewees contested this title, opposing it to the idea of conservative and slow-in-change city.

47 The Mind Bomb project was started in 2002 by a group of artists, architects, journalists and writers as a pioneering action. It was the introduction of the socio-political poster in Romanian culture and "became a means to hack into the dominant discourse of mainstream politics, the mass media and the advertising industry. The ultimate purpose of these actions is the creation of autonomous space for the purpose of dialogue and debate that initiates social change. Mind Bomb started as a challenge to the consumer culture and later developed into a broader critique of contemporary Romanian society and its public space. ", as they assume it on http://www.mindbomb.ro/.

their performances on the peripheries of Cluj, in Mănăştur neighborhood and also in one of the malls of the city, also located in a former communist workers' outskirts, which has recently become a new commercial and symbolical center for many people from Cluj.

Under these circumstances, the idea of repetition was the leitmotiv of the whole project and invited reflection on daily practices in the urbanscapes.

The curator of the project claimed:

> Based on this concept of taking over and reflexion through the artistic act of certain repetitive social gestures, within a completely different approach of the public space: the area of Mănăştur neighborhood, a space charged with themes like urban density, playgrounds for children, the space of neighborhood meeting, the pedestrian and road traffic, etc. (IS, 43, M)

[...] "Samples" was an exercise of modeling the urban context through repetitive events. Starting from the daily little rituals of the individuals to the group behaviour, the project was processing thematically recurrent activities of the urban conglomerate's inhabitants, giving a new angle to the social life of the area and its people and questioning its own actions. Artists and dancers developed a performative model of the urban context, based on audio, video and human behaviour patterns. They took from the urban context recurrent sounds – the noise of the street, of the cars, of the steps, images of certain repetitive visual patterns – windows, blocks, the dynamics of the traffic flux and the standard behaviours of the inhabitants of the city. This project was an integrated show of sound, image and movement, with the direct and immediate public participation. (IS, 43, M)

I focused more on Mănăştur's case, due to the fact that the mall site lacked the attention of the 'clients' and did not raise too much attention. The innovation brought by "Samples" in Mănăştur was that the periphery became of special significance to a city in which art, whether focused on galleries or urban space, is mainly concentrated on (city) center-based popular themes and is still under centralized operation.[48]

---

48 Counterposed with this, I have been recently noticing another tendency regarding the central square in Cluj this time, which became for me the "Screen Square" of the city - accommodating TIFF (Transylvania International Film Festival). It has 'conquered' this place and created a precedent of this type of space usage. After the TIFF events in June 2010, the Town Hall gave usage permission mostly to private actors for screening projects attracting mass public attention. The square provides interesting data from that analysis of the simultaneous processes of the city public space culturally developed and at the same time commercialized.

Initiating a project like this, within the working neighborhoods of Cluj, these cultural actors had created a series of, for and against reactions. But the most important thing is that it brought 'the art to the people', and not only to the upper class. It was a type of art created both within and for the context where it was performed. It was also 'site-specific', having inner reasons of activating spaces which would otherwise have been only used for certain predefined purposes and daily practices that have not been changed in the last forty years or more, creating an urban reflex of usage across an even bigger timeline. The neighborhood's reaction was greater than expected[49]; it regarded the event as something that had never happened to the community before in those places. Local people used this location in daily life just for transit, going to or returning from their jobs. The physicality itself of this passage (in Romanian called "Pasajul Minerva", Minerva being the name of this area within the neighborhood) makes this place, a transit space *par excellence*, a bridge between the residential blocks and the commercial ones, now represented by a McDonalds, some small shopping sites and one of the main tram and bus stations in the area. For this reason, the organizers chose the space to first analyze and then suddenly disrupt a daily flux, causing reflection, interaction and then sharing with the community an art experience. The idea of public participation was another issue strongly brought into consideration.

In comparison, the event organized in the Mall had a different output: people there did not pay too much attention to the performance and mainly continued hurrying for last minute shopping as it is usual in that space. But it also proved that a different type of space, for artists trying to adapt their performances, can be a challenge for the cultural actors. Cluj individuals seemed to be more accustomed to (and interested in, within this environment) the particular commercial events typically offered by the malls, like culinary offerings and other types of fairs or children programs, than to the dancers and media artists from "Samples" who attempted to break the space into new usages by playing with its elements in order to deconstruct its commercial role and to propose, through videos, a critical discourse towards consumption.

In brief, what these two "Samples" case studies present is first, how few projects are being proposed to different urban communities, to provoke them into sharing the urban space in a new way. In a city where mainstream culture is still by far the dominantly-used mean for 'entertaining' people, these few cases are trying to appropriate certain areas of the neighborhoods and to give impulse

---

49 Measured especially in interactivity indicators and also in after-event reactions recorded on the site and through interviews realized after the intervention.

and voice to new actions based on their needs and concerns about their places in the city.

Compared with the cases from Bucharest, we can state that we have the story of a smaller but more creative city in which local initiatives can prove to be more efficient at this level. Moreover, the Bucharest scene is already offering other types of events in public space (some of them contested), which little by little are reshaping the public attitude towards art and city matters. Meanwhile the Cluj scene is trying to show its public what greater power public performance has, as a new tool of urban reflexivity.

Instead of a conclusion

Artistic manifestations, similar to other forms of temporary usage of public space, have the capacity to highlight opportunities the urban space offers and which are, most of the time, ignored in development strategies. The most important cited goal of integrating these manifestations into the public space is linked to the rehabilitation of the disadvantaged urban areas.

When speaking of the general overview of Romanian artistic manifestations in the public space, CCCDC (2009) assumed that:

> the social message, until now, is practically nonexistent in the actions promoted by the local authorities. The artistic interventions organized by the non-governmental associations or the private actors are still the minority and rarely function as a critique of the city, lacking visibility in the native space, even if remarked on the European level. The artistic events with a social catalyst role are practically nonexistent, (...) where the actions like performances are predominantly conceived in a commercial logic, the mass success of the events being more important and not their social message (pp. 12–14).

And in speaking of the Bucharest public, CCCDC prefers to characterise it as a "passive and exclusive receiver". But regarding this type of public, I would say that this is sometimes more a symptom of living in capital cities, in which the city flux often ensures its inhabitants ignore certain interventions and artistic actions, while letting them continue their work- and consumption-focused daily urban paths. Nevertheless, although the public of Romanian cities still has a lot of patterns imposed on its behaviour, and 'gates to its heart and soul' often stay closed to such art interventions, there still exists good potential for involving them into future reflexivities. The cases presented above show that where there is cause there is also an effect.

As long as the urban public space suffers its own illnesses due to the problems of the whole body of city experiences, cultural actors will have the capacity to 'inject' into it certain cures offered through art. But we mustn't take all cures as inherently beneficial. Geenens and Tinnevelt (2009) argued for "physicality" in the course of performing democratic roles between the actors engaged, "because it strengthens interpersonal communication and makes it 'more civil' and because it makes 'the ethos claims' easier. Furthermore, they believed that "because physical performance has an 'impressive' effect both on showing leaders that large number of people care about an issue and by impressing the seriousness of binding collective decision making on participants" (pp. 108–112).

Art and culture could have a strong influence on decision-making if politics were more careful considering them. Culture is represented also by indicators of society's symptoms which they show in a 'triple-sided mirror': remembering the past, illustrating the present status of the problems and envisioning future changes that could happen if we listen to the prescriptions. But another risk is the exclusivity of the domain that can often turn into only caring about the artistic community itself. In this way, there is a risk of self-isolation and breaking connection with society and the local community. Public space is the ideal arena for expressing a message that has to be quite clear for the given conditions. "The message still matters", as Geenens and Tinnevelt (2009) remind us when speaking about democracy and public space.

In my opinion, there is also a need to integrate artistic events in the public space with the politics of urban development and with local economical strategies. Independent cultural actors are creating interventions that are still reactions to the state actions upon the local public space.

At a public debate, architects like Gheorghe Pătrașcu admitted that "the public space, at least in the case of Bucharest, has all the chances of getting into the clique of the vanishing species" and also that "we do better in theory than in practice concerning this topic" (Matzal, 2009). However, there exists a minority of actors who are working on the future basis of rebuilding civil society or at least building a constant practice of public space events that could reach beyond the theoretical frame and really affirm themselves in the public sphere. To be truly successful, they must not impose themselves but rather contribute to the 'bottom-up' process of cultivating public art as rooted in the public mind and gathering together artists within local communities to contribute to these dynamic movements.

References:

BABIAS, M (2008) Spatiul Public Bucuresti/Public Art Bucharest 2007: Idea, Design, Print. Cluj, 2008

BUDEN, B. (2003) Public space as translation process. Available from: www.republicart.net/disc/realpublicspaces/buden03_en.htm, pp. 3–5.

CENTER FOR STUDIES AND CONSULTING IN THE CULTURAL DOMAIN BUCHAREST (CCCDC) (2009) Culture in public space.The analysis of the artistic events taken place in Bucharest. Available from: http://www.culturadata.ro/, pp. 10–20.

CERTEAU, Michel de (1984) The Practice of Everyday Life. University of California Press, Berkeley, pp. 37–38.

GEENENS, R. and TINNEVELT, R. (2009) Does Truth Matter? Democracy and public space. Springer, Leuven, pp. 108–112.

MASSEY, D. (2007) For Space. Sage Publications.

MILES, M. (1997) Art, space and the city. Public art and urban future. Routledge, London, pp. 52–113.

MIHALI, C. (coord.) (2005) Art, technology and public space. Paideia, Bucharest, pp. 99–100.

CHAPTER 5

Contemporary Art: Between Action and Work. The cases of Krzysztof Wodiczcko and Jenny Holzer

Celia Ghyka

> "Man's dignity demands that he be seen (every single one of us) in his particularity and, as such, be seen – but without any comparison and independent of time – as reflecting mankind in general."
> *Hannah Arendt, Lectures on Kant's Political Philosophy*

Introduction

While public space has been described by much recent scholarship as declining, we are now witnessing a surprising proliferation of spatial and political practices that are mainly pretending to restore it. These practices attest to a growing interest in the issue of public space within the discourses of urban policies: even when these discourses are no more than empty rhetoric, the mere presence of interest can be seen as a symptom of a change in perspective for the public agenda. These practices also demonstrate a shift in artistic practices involved with the construction of physical and urban public space towards critical issues such as resistance to the dominant aesthetic that is primarily imposed by the language of the media and advertising.

These critical practices are moreover defined by a tendency to broaden or challenge the artistic experience through a range of alternative mediums, such as performance, video, social art, body art, and installation. The de-definition of art (Rosenberg, 1993) and the loss of medium specificity in recent modernity has often been discussed and analyzed[50]; but it is not the intention of this article to investigate the genealogy of these artistic transformations.

The intention here instead is to discuss the relationship between contemporary artistic practices and the construction of public space in the city. The starting point will be the influential theory of Hannah Arendt that holds public space to be essentially political and defined by action as its key feature. The above-mentioned observation about medium specificity is therefore one of the main arguments for a rereading of Arendt through the lens of contemporary

---

50 See Krauss, R. (2010). Perpetual Inventory. Cambridge, Massachusetts: The MIT Press or Rosenberg, H. (1993) The De-definition of Art. Chicago: University of Chicago Press, for excellent accounts of these transformations.

art practices. The question is whether this contemporary context of art, characterized mainly by their temporality and political agenda, can provide a starting point for a reconsideration of a conflict-prone open public space.

Arendt's book, *The Human Condition*, first published in 1958, has been widely influential. A major part of its influence grew from within the artistic and architectural education milieu[51] of the '70s, first in America and then later in Europe. In this book, Arendt describes three fundamental activities of men: labor, work and action. These three activities characterize for her an essential part of human life that is *vita activa*, underlying all human activities. Labor serves to maintain sheer physical existence and "corresponds to the biological process of the human body", involving activities in which the results are consumed immediately (such as cooking, cleaning, maintenance). Work, on the other hand, is the activity that provides "an artificial world of things", manufactured objects that remain in the world for some time. And finally there is action, "the only activity that goes on directly between men without the intermediary of things or matter" (Arendt, 1999, p. 23). This concept of action produces a communicative space which for Arendt constitutes the political sphere par excellence and is the condition for public space to come into being as a political sphere.

The intention of this article is twofold. I aim to reposition Arendt's perspective on the distinction between action and work onto the scene of norm-challenging contemporary art practices. I also will utilize Arendt's conception of the public realm as being essentially a realm of action to explore how contemporary art works create or question urban public space.

In order to illustrate the double movement between a work of art and public space as understood through Arendt, this article will first turn to the paradigmatic contemporary art practices of Krzysztof Wodiczko and Jenny Holzer. Both artists use architecture and urban public space as a stage and support for their art interventions, which consist primarily of moving images projected onto significant buildings. Wodiczko mainly focuses on projecting the history of others (usually those that are socially invisible) and thus offering them visibility, whereas Holzer uses language – publicly printed words – as a means to raise consciousness about hidden violence in our society. These projected images challenge the image of "ideal" public space as being both potentially harmonious (Habermas, 1990) and opposed to a domain that is permanently exposed to conflict and negotiation (Deutsche, 1996). In the second part of the chapter, I will juxtapose these artistic efforts aimed at the re-construction of urban public space with Arendt's strict delimitations of action and work, and will address the issue of the existence of new practices in and about the public space.

---

51 Such as Kenneth Frampton (1979) or later George Baird (1995).

Wodiczko: the space of appearance

Krzysztof Wodiczko (born in 1943, Warsaw, Poland) is an American artist of Polish origin. At the end of the '70s, Wodiczko was experimenting in Warsaw with prototypes of something the critic Tom Finkerlpearl has named "subject-oriented machines". These were the first examples of a series he would later call 'critical vehicles', and on which he would be working for several decades.

Among his schematics were those for a café-vehicle hybrid, a kind of moving table and chairs mechanically activated by the intensity of the discussion and gestures of the persons who occupy them. In other words, the machine moves only when people either talk or communicate through gestures and the speed depends on the intensity of the communication. These vehicles obviously offer an interesting way of approaching the question of public space. They serve as metaphor and as an ironic interpretation of the origin of the public sphere as placed by Habermas in the coffee houses of 18th and 19th century European cities.[52] Wodiczko also speaks of the stimulating effect of coffee that is, in his opinion, related to the necessity of continual invention in public space (in order to keep it alive). Although the machine is very carefully designed in the drawing, it is not meant by its author to be actually constructed. Its conception on paper posits public space as being conditioned by speech. As long as debate is going on, society is moving. When debate stops, everything stops:

> "The Vehicle-Cafés are more of a metaphor. They work on a representational level; they are not to be enacted. The Vehicle-Café 1 could be operated by contemporary intellectuals as it was operated by the intelligentsia in 1970s Poland. It could be an allegory of whatever is still left of our interest in existential philosophy. That is, questioning to what degree we are collaborating in the creation of these kinds of machines for ourselves, to what degree we are forced to be passengers, as imaginary drivers. Another kind of vehicle is being created now as we sit here and discuss these Vehicle-Cafés. This is a New York coffee shop, and we continue their conversation." (Wodiczcko, 2003, p. 71)

Microphones installed in the headrest are connected to the engine. As long as the operators continue to speak – regardless of the content – the progress of the machine is guaranteed. The operator's movement induces an axial movement of the table and chairs of the vehicle. This movement is then transmitted to the wheels through a gear. The vehicle can only move forward and backward, and

---

52 Just as ironically, Wodiczko evokes the supposedly Polish origin of the coffee-houses in Europe, related to a Polish spy in the battle against the Turks who asked to be paid with the coffee left back by the Turkish army. With those supplies he opened the first coffee house in Vienna and thus originated European public space. Wodiczko, K. (2003).

as long as the speakers are gesticulating the machine's progress is guaranteed. The metaphorical discourse can here be seen as a game of mirrors: we must not forget the origin of the word metaphor, which in Greek means the transport of a thing from one place to another. As Michel de Certeau (1990) reminds us, in contemporary Athens, public means of transportation are called *metaforaï*. Wodiczko's vehicles act as a play of *mise-en-abîme*: they are metaphors that are literally intended to "transport" their passengers.

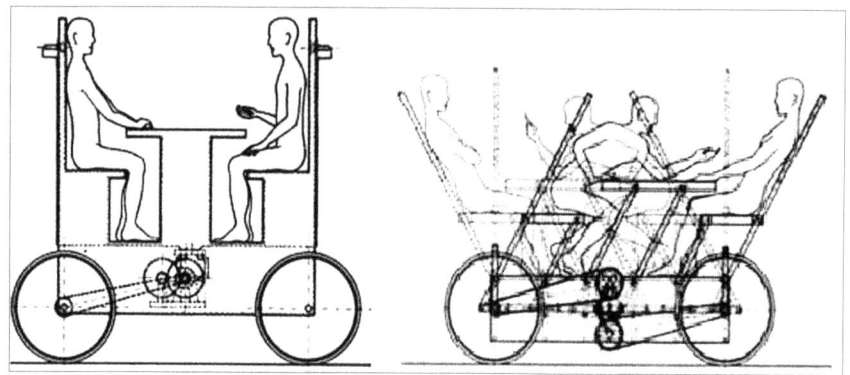

Picture 1: The Vehicle-Cafés.
Source:A Dean & DeLuca Coffee Shop Conversation
with Krzysztof Wodiczko. Perspecta 34

After the vehicles period, Wodiczko began experimenting with large outdoor projections on buildings as the main medium of his art. These projections make monuments actually seem to speak, telling untold stories of individual suffering or hidden violence. In the space of the projection, the artist gives people the opportunity to speak about their traumatic experiences. His approach is not only critical but also clinical.

This method is used with particularly arresting effects in two of Wodiczko's projects: the Hiroshima Projection and the Tijuana Projection, both of which address the public in an interactive way using aesthetics (as fundamental to artistic choice) to participate in the construction of public space.

The Hiroshima Projection was performed in 1999, and focused on the A-Bomb Dome, one of the very few buildings that survived the bombing of 1945. For over half a century now, the desolate structure has been maintained as a ruin, in order to serve as a memorial.

Wodiczko's project consisted in projecting the voices and gestures of contemporary inhabitants of Hiroshima onto the empty skeleton of the Dome.

The movements of the speakers' hands animated the facade while they spoke, enlivening the empty structure. The monument thus became a depository of past and present memories; anthropomorphised, it turned into a faceless, uncanny body that spoke of human trauma:

> I started working on my Hiroshima projection with the assumption that we were going to 're-actualize' the A-Bomb Dome monument and reanimate it with the voices and gestures of present-day Hiroshima inhabitants from various generations, starting with those who survived the bombing, who witnessed it; their children, who may still remember; their grandchildren and great-grandchildren. So all those generations somehow connect through this projection, not necessarily in agreement in terms of the way the bombing is important and the way the meaning of that bombing connects with their present experience. The fallout of the bombing is physical and cultural, psychological.(Wodiczko, 2005)

Wodiczko started by interviewing the survivors of the Hiroshima bombing, not only Japanese but also Koreans who were at that time slave-laborers in the city and whose voices had not been heard as equals in the official memory. The project gave these former workers a public voice for perhaps the first time. Younger generations of Hiroshima inhabitants participating spoke of their silent traumatic domestic experiences as if they were bearing some genetic stigmata. The artist states: "it is sometimes easier to be honest speaking to thousands of people through a monument than to tell the truth at home to the closest person."(Wodiczko, 2005)

Here architecture is more than a projection screen: the building is engaged in the airing of memories, mainly traumatic ones that have been neglected and/or are invisible to the official culture. Thus the surface of architecture which the artist calls the 'skin of the building' becomes a texture interposing itself between the person confessing (individual memory) and the public (a receiver and, in turn, a participant in the act of creation of a new collective memory). The monument itself also serves a silent witness, making public confession possible by acting as a transference medium. The well-known ruin is able to validate personal stories in its role as a silent witness and then reflect them to the public. The physical presence of the building as the site of broadcast circumvents the direct relationship between individual sufferance and the public.

Commenting on the Hiroshima Projection, Rosalyn Deutsche (1996) introduces the idea of "public vision", drawing on Emmanuel Lévinas' thesis on the apparition of the Other. In her opinion, Wodiczko challenges the way in which contemporary advertising media typically uses faces for "triumphalist purposes", whereby the Other is denied as a real presence, lacking individual history and personality (this, of course, in an idealized world where reality must adequately equal its media-produced image). Wodiczcko's work acts in the realm of the visible in a double way: it offers visibility to both the others and the artist himself. Fundamentally he proposes a non-indifferent way of seeing, one where participation and sympathy/empathy (the ability to put oneself in the place of the other) is required. The construction of public space is, in this interpretation, intimately linked to this way of seeing and to a sort of critical vision that implies the responsibility of the one who sees.

For the Tijuana Projection (2001), Wodiczko reverses the roles of the monument and of the participant – this time faces literally cover and substitute for a monument. As he states, "the skin of the building and the skin of the person are shifting focus". His sketches are very interesting in this respect: they show clearly how he understands the operation of transfer between face/voice /story/architecture.

The Tijuana Projection consisted of interviews with women who have suffered violence (mainly sexual abuse) and been too frightened or traumatized to speak openly about it. They told their stories in interviews conducted directly in front of the building of the Tijuana Cultural Center (representing quite literally the official local culture). The projection resulted in their faces becoming the skin of the building, warped as needed to fit the shape of the 'speaking' architecture.

The highly emotional way in which Wodiczko uses these traumatic memories may raise questions concerning the resemblances between these projects and reality-TV shows, and the ways trauma can be instrumentalised in order to be exposed. Wodiczko's approach can also be read as a critique of this manner of packaged suffering.

The artist uses the methods of reality-TV shows and of media-advertising while clearly subverting their aims. To him the projects not only contribute to transforming urban public space into critical space, but also work towards the achievement of a different and therapeutic aim: "The process of unlocking the post-traumatic silence requires not only critical, but also clinical, approaches and attention." (Wodiczcko, 2004). His projects subvert both the use of public buildings as a support for advertising and the use of confession in order to increase ratings for reality-shows. We are used to seeing large images on buildings advertising images of a perfect life, attempting to contribute to our own perception of reality and of the city. These images almost always appeal to the rhetoric of happiness. In contrast, Wodiczko's animated buildings speak of traumas that require witnessing and mourning to be surpassed and, as in psychoanalysis, this work always passes through words.

In addition, the fact that the traumatic narratives take over the bodies of public monuments is a way to engage monuments (usually static and non-responsive) with the real-life of those who inhabit or see them. Engaging urban public monuments with what is happening today is a way of rendering them responsive to the present-day collective concerns, fears, and sensibility. This prevents them from remaining merely linked to a heroic history and makes them re-act to the present.

Wodiczko's projections, therefore, make the city buildings monuments of the histories of outsiders, the vanquished, the victims of the official history, and offer them a bulwark against oblivion:

> I have employed urban sites that are charged with some meaning, representing something that Walter Benjamin called the 'history of the victors', or one might say, the 'culture of the victors'. The sites are monuments or other structures, and now the city at large. Into this, I insert what Benjamin calls 'the secret tradition of the vanquished' (Wodiczko, 2001, p. 339).

Offering public visibility to gesture and speech brings us back to the definition Hannah Arendt gives to political public space as appearance in the widest sense of the word, namely, the space "where I appear to others as others appear to me, where men exist not merely like other living or inanimate things but make their appearance explicitly". (Arendt, 1999, p. 50).

For Arendt, public space rises "directly out of acting together, the sharing of words and deeds", leading us to address the work of an artist who uses words as her medium, bringing them directly to the public space of the cities.

Jenny Holzer: writing in the public realm

Jenny Holzer (born in 1950, Gallipolis, Ohio) is an American conceptual artist. Though she initially started as an abstract artist, she began working with text as a central component of her medium at the end of the '70s. The main focus of Jenny Holzer's work is the use of written text in the public realm. Street posters are her favorite medium, but she also makes use of a variety of other media, including LED signs, plaques, benches, stickers, T-shirts, and the Internet. She has performed projections on buildings, mountains and rivers throughout Europe and North and South America.

As in the case of the popular artists Christo and Jeanne-Claude[53] and their large, public, and less confrontational installations, Holzer deals with reclaiming

---

53 American artists Christo (Vladimirov Javacheff, Bulgarian born) and Jeanne-Claude (Denat de Guillebon, French-born) are known for large-scale environmental works (both urban and landscape), where they mix mediums such as painting, sculpture, architecture, installation and urban planning. Their most famous works include the wrapping of the Pont Neuf in Paris (1975-85) and the contested wrapping of the

urban space. She uses projected letters and words that move like a kinetic light sculpture in public space. Words glide along architectural surfaces and then enter the ground; they superimpose themselves as a second skin of a building. The effect of these moving writings sometimes resembles the credits rolling at the end of a motion picture or of a TV show, thus reaffirming the close connection with cinema and TV as a medium. And again, using advertising's technique of making words visible in the public realm, Holzer addresses public and political issues, thereby subverting the typical aim of advertising. As the artist states:

> I show what I can with words in light and motion in a chosen place, and when I envelop the time needed, the space around, the noise, smells, the people looking at one another and everything before them, I have given what I know. (Holzer, 1998)

The public dimension is integral to Holzer's work, and for a decade, the open display of words has been a critical component of her multi-disciplinary practice. Holzer uses poetry too as part of her artistic medium, which is concentrated on large light (thus immaterial) sculptures projected onto public buildings. In some of her projections, like the London Projection from 2006, she proposes an interesting fusion between light, cityscape or urban space, and poetry. For these projections she used poems by the Nobel Prize-winning Polish poet Wislawa Szymborska. We may identify here some obvious influences coming from the Dada technique of the destruction of language and research on words in public space, as Dada artist Hugo Ball stated: "The Word, gentlemen! The word, so to speak, is a public concern of the first importance!" (Louis 2006, p. 38). For other projects, like the Xenon Projections[54] (which she has been making in cities all over the world, including London, New York, Sienna, Turin, Liverpool, San Diego, Venice, Vienna, and Singapore) the poetic quality is intrinsic to the performance itself, architecture being literally dematerialized by words that interrupt the building and impose themselves into the public space.

Words and signs are not alien to the urban architecture. Robert Venturi, Denise, Scott Brown and Steven Izenour note in their essay on Las Vegas (Venturi et al., 1977) that communication dominates space as an element in the architecture and the landscape. As the critic David Joselit (1998) has pointed

---

Reichstag in Berlin, a work where the negotiations with the German Bundestag lasted for 24 years (1971-1995) until the final approval was given. Their works take possession and completely reconfigure urban space through the means of ephemeral installations (mostly giant fabric wraps). http://www.christojeanneclaude.net.

54   For an exhaustive illustration of the Xenon Projections, visit the artist's site: http://www.jennyholzer.com.

out, language is often assumed to be deprived of materiality (without weight or dimension) and appears to be an ideal means of abstracting. Yet the contemporary city is largely constructed with words and signs that dominate and describe urban spaces, as well as transitional spaces of communication and exchange. We could hardly imagine an environment without signs nowadays. Legible signs and texts are also dominant in the "non-places of the super-modernity" theorized by Marc Augé (1992) – places of anonymous transportation, consumption and entertainment that define our daily built environment. According to Augé, these places "have the peculiarity that they are defined partly by the words and texts they offer us" (Augé, 1992, p. 102). The architectural character of these places is largely modified through these written signs that establish their functional legibility (the instructions), as well as transforming them at the same time into commercial advertisers that address a mobile spectator. Along with solid architecture and spatial environment, text becomes a spatial definer.

Unlike Wodiczko's projections, where words are spoken out loud in public through the 'body' of the building, Holzer's method is closer to advertising techniques and it is not a coincidence. She challenges the ways in which written text's omnipresence in our daily urban environment transforms it into an almost invisible reality whose workings we cease to question. Modern urban space has taken its contemporary form also through its commercialisation, and Holzer exploits modern billboards for critical purposes, re-appropriating the methods and techniques of commercial pop-culture through a critical and ironic use. She uses urban space as a stage for her work, where written words are intended to re-define architectural spaces.

Wherever the viewer is prepared to see commercial messages – large LED or TV screens on buildings – Holzer inserts critical or poetical statements that reverse the function of the advertisement. Text thus ceases to be a way to sell a product and becomes a way of raising awareness on public issues or takes the form of a poetic statement trying to make the viewer stop and think. In an ironic play on the hypocritical language used by advertising to persuade potential clients of their own interest in buying a product, she confronts the public with statements such as "It is in your self-interest to be very tender", "Savor kindness because cruelty is always possible later", or "Protect me from what I want".

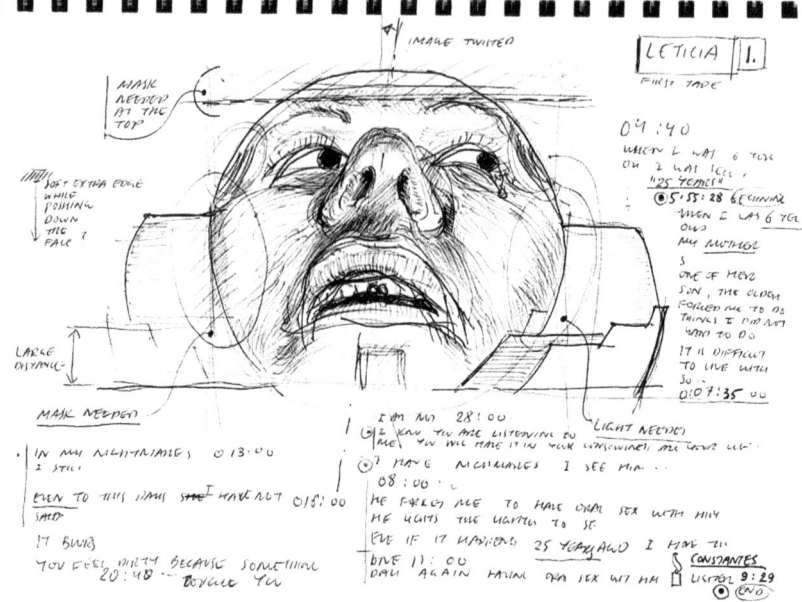

Picture 2: "The Tijuana Projection". Public projection at the Centro Cultural de Tijuana, Mexico, 2001. © Krzysztof Wodiczko.
(Source: Public broadcasting service, http://www.pbs.org/art21/images/ krzysztof-wodiczko/the-tijuana-projection-2001-0?slideshow=1)

These works were exhibited in large audience urban spaces such as Piccadilly Circus in London and Times Square in New York.

By using a cinematic technique, Jenny Holzer is also challenging the illusionism of the commercial media, using it for critical aims (political consensus, hidden violence, etc.) in order to dismiss illusion. Her work reminds the viewer that no text, including advertisements or government announcements, is ever deprived of a place, a time, and an author.

We may nevertheless question a certain sense of ambiguity she introduces in the words she uses; words can be difficult to decode and fragments of text can be sometimes interpreted in opposite ways. Also, one may question the spectacular or too obvious way of using visual montages as a means to dematerialise urban contexts.

Another critique that has been addressed to both Wodiczko and Holzer's work is that they deliberately use social or public issues in order to gain visibility as artists. Yet visibility is one of the essential conditions of the artist: an invisible artist does not exist, especially if his/her main concern is urban and public space. The artwork reveals the Other as a 'subject', but it also reveals the artist, providing both with a visibility that is welcomed by the viewers (we addressed this issue already in concern to the projections of Wodiczko). An important issue raised by works such as these addresses the very space of appearance, asking about the experience of being public and who has the right to appear, and be exposed to the appearance of others, in public space. This condition of visibility upon which reality itself depends, brings us back to the starting point, that of Arendt's conception of the realm of appearance.

Work of art *and* action

The visibility of a person – a visibility for "every single one of us in his particularity" – is for Arendt a virtue. It is visibility that guarantees the publicity of our deeds and of ourselves and thus gives us a chance to overcome oblivion and mortality. Visibility is a precondition of appearance, of "something that is being seen and heard by others as well as by ourselves" (Arendt, 1999, p. 50), something that constitutes a person's being and reality, something that exists thanks to action and speech.

Not only is the public realm the space that gives us visibility, Arendt tells us, but this space should give visibility to each of us – "every single one of us in his particularity" – and the absence of this possibility equates a denial of human dignity and freedom, almost a fall into slavery. It is visibility that guarantees the publicity of our deeds and of ourselves and thus gives us a chance to overcome oblivion and mortality:

> many ages before us men entered the public realm because they wanted something of their own or something they had in common with others to be more permanent than their earthly lives. Thus, the curse of slavery consisted not only in being deprived of freedom and of visibility, but also in the 'fear of these obscure people themselves that from being obscure they should pass away leaving no trace that they have existed'. (Arendt, 1999, p. 55)

But more than a space that should leave a place for "every single one of us", "the reality of the public realm relies on the simultaneous presence of innumerable perspectives and aspects in which the common world presents itself":

only where things can be seen by many in a variety of aspects without changing their identity, so that those who are gathered around them know they see sameness in utter diversity, can worldly reality truly and reliably appear. (Arendt, 1999, p. 50)

Thus, a space is posited that is potentially participatory (every one can appear and be seen) and elastic (offering multiple points of view of the same thing) – conditions that may allow us to speak of Arendt's definition as an open definition of public space.

In Arendt's terms, the public space is understood as the "space of appearance", one that "comes into being whenever men are together in the manner of speech and action", and this space of appearance is always, in her opinion, a potential one. This place therefore exists to be filled and fulfilled through action and speech, continuously evolving in order to make space for the next action and speech.

The characteristics of the public realm appear more clearly in the distinction she proposes between work (as a condition of the worldliness, defined by its singularity) and action (as a condition of plurality).

Action is a condition of human plurality that reveals itself through speech, for "no other human performance requires speech to the same extent as action" (Arendt, 1999, p. 179). More than just a means of communication, in acting and speaking (as in all other performances), people show who they are, and moreover, they reveal it to others, appearing and being heard. For Arendt, this "revelatory quality of speech and action comes to the fore where people are with others and neither for nor against them – that is, in sheer human togetherness." (Arendt, 1999, p.168). Thus, action would be the paradigm for the political as an activity that only makes sense through plurality, for

> it is only action that cannot even be imagined outside the society of men.... Action alone is the exclusive prerogative of man, and only action is entirely dependent upon the constant presence of others. (Arendt, 1999, p. 23)

While defining work in general as an "activity which provides an artificial world of things", Arendt reserves nevertheless a very privileged place for the work of art. The work of *homo faber*, as she puts it, consists in reification (Arendt, 1999, p.180), characterised by solidity and the material character taken from being produced by human hands. Works of art occupy a privileged place among work in general, as they are "objects which are strictly without any utility whatsoever", objects that must not be used but instead carefully removed from the context of ordinary usage so as to attain their place in the world. They are permanent and durable and thus reveal the durability of the world itself, as a "mortal home for non-mortal beings"; "because of their outstanding permanence, works of art are the most intensely worldly of all tangible things". Arendt grants privilege to the very particular role and place that works of art occupy among things:

> it is as though worldly stability had become transparent in the permanence of art, so that a premonition of immortality ... has become tangibly present, to shine and to be seen, to sound and to be heard, to speak and to be read." (Arendt, 1999, p. 168)

Even if reification for works of art consists in transfiguration (of acts and deeds and speech) as "thought things", this does not prevent them from "being things", characterised by permanence and the act making (fabrica).

For Arendt, one of the characteristics of action is to leave no trace, and that is why acting and speaking people need the help of the artist in order to make their action survive. If works of art are characterised mainly by their "outstanding permanence" and their tangible nature, action's main characteristic would be its impermanence, as in speaking.

But among all works of art, Arendt gives a special space to poetry, and this is due to its lack of materiality. "Of all things of thought, poetry is closest to thought, and a poem is less a thing than any other work of art". Yet, she continues, "even a poem, will eventually be 'made', that is, written down and transformed into a tangible thing among things, because remembrance and the gift of recollection need tangible things to remind them." (Arendt, 1999, p. 170)

One may go further and say that poetry occupies a sort of intermediary position on the border between action (impermanent and defined by speech) and work – a tangible thing among other things produced by men. Such a shift can be illustrated by Jenny Holzer's cinematic sculptures that make poetry briefly visible in the public space. This play between the permanence of the written word in poetry (as in a book, for Arendt) and its ephemeral appearance in Holzer's projections, might point to a new understanding of the relationship between work (of art), word (action), and poetry (as a border-object between action and work). What Holzer proposes is to bring the poem, which has been written down and thus transformed into a tangible thing, into the public realm while providing it with the impermanence of a public appearance, giving it back, so to speak, its ephemeral qualities of thought and action.

Two of the main critiques that have addressed Arendt's theory refer to her disjunction of the social and the political and to her aestheticised view of the political, idealised through the image of the classical Greek cities.

As Seyla Benhabib (1990) has shown, we may distinguish between two kinds of public space as defined by Arendt: the agonistic and the discursive. The space of appearance that defines the public realm, where moral and political qualities are revealed, is seen essentially as a space where one competes for recognition and acclaim, seeking a "guarantee against the futility of... individual life" (Arendt, 1999, p. 55). This is the ideal public realm as Arendt sees it – originating in the Greek cities, where heroes reveal themselves in the whole beauty of their heroic action. It is this aspect that has been often criticized as an aestheticisation of politics. Within this logic, the public space itself, being essentially the space of shared politics, could be in its turn aestheticised as well.

On the other hand, the discursive approach to public space situates its emergence whenever people act together in concert, thus becoming the space where "freedom can appear" (Arendt, 1993, p. 4). What qualifies this space as public would then be the presence of common action coordinated through speech and persuasion. For Benhabib, this second type of discursive space would characterize the modern public realm, rejoining the classical Habermasian definition of the public sphere. Important in this distinction is the porosity of the public realm, its consequence being a less strong (or even pointless) distinction between the social and the political in the contemporary world. If for Arendt the emergence of the social within the political corrupts the political, as Benhabib states, this distinction has nowadays become less relevant, primarily because the struggle to make something public has become a struggle for justice. Thus every group that struggles for its own justice extends the scope of the public realm: women, family and the private sphere become political issues. Therefore the topics for public and political conversation can no longer be pre-defined by an agenda, for the very definition of this agenda would equate with the instauration of a public sphere.

The criticisms of Arendt's view on politics as being deeply aestheticized[55] are often linked to agonistic qualities. But as Benhabib has shown, the agonistic is not the only quality of publicness as defined by Arendt. Moreover, the public space as the space of appearance – where every one of us can see and be seen – obviously has an aesthetic character, that is inherent to our perception of reality. As Jacques Rancière (2000) has suggested, modernity is an "aesthetic age" (régime esthétique), defined mainly by a new sensibility through which art can be identified as opposed to a specific type of praxis (making). This new type of sensibility implies an attention to life itself as being ruled by the same laws as art, an attention that is itself of aesthetic nature.

---

55 Researchers as Martin Jay argue that the aestheticization of politics is for Arendt the disjuncture between the political and any rational, utilitarian or social foundation. In the same manner, George Kateb explores parallels between Arendt's work and fascist thought. For a large discussion on critics of Arendt's aesthetics, see Kimberley Curtis (1999, pp. 17–18).

Arendt's aesthetic has deeper[56] grounds than the mere formal appearance on the public stage, where aesthetics is situated at the very foundations of politics. To her, aesthetics are a quality that enhances our sense of reality through the experience of appearance.

This quality has been emphasized by Dana R. Villa (1999) when speaking of Arendt's theatrical vision of the public realm as a "stage for words and deeds". According to Villa, Arendt's point is that a vibrant sense of the public tends to be found only in those cultures where a form of social theatricality is clearly present and manifests itself as a second nature.

An excellent (albeit critical) account of the relationship between politics and aesthetic in Arendt's work is given by Kimberly Curtis (1999). She states that we need a reading of Arendt "that takes aesthetic experience not as an analogy for an account of political life, but as a central feature of it as a space of appearances". She shows how for Arendt, aesthetic perception is integral to our openness to reality. According to Curtis, Arendt places an enormous ontological and existential burden on the public sphere as a space of appearances.

In particular, it is this one domain in which freedom can become a tangible experience. As such, the public sphere is the site for a miracle-like capacity of humans, in responsive provocation to begin a new course of action and thus stand out against a world whose profound inertia always threatens their freedom. Further, it is in the public sphere that our sense of the world as a place, shared and with a certain temporal continuity is perpetually won and renewed. In short, the public sphere is the fertile ground in which a sense of human reality and meaningfulness can grow. (Curtis, 1999, p. 21)

But more than the sensibility or openness to the Other, human freedom is intimately linked to the possibility of appearance, and inversely, being deprived of this possibility is seen by Arendt as an insult and a form of injustice. Curtis develops this relationship by referring to oblivion as an insult, more unforgivable than any other.

We may now recall Wodiczko's public projections and the way he proposes non-indifferent ways of seeing in the Hiroshima and Tijuana projections, as well as in his other work. The artist is concerned with offering a public voice and appearance to those who have formerly been deprived of it, thus repairing the insult of oblivion. His work only exists if people gather together as spectators who enable the visibility of those that speak in public and activate the

---

56 I refer here to the Kantian tradition that grounds aesthetic in the common sense. It is particularly through aesthetic taste that Kant discovers the sensus communis, that he describes as a transcendental idea, one that cannot be proven in reality but exists nevertheless, and Kant admits its existence through the fact that we judge aesthetically.

spectatorship. In the same way, Holzer uses projections of words that engage the spectator with critical issues, forcing a non-indifferent response.

As we have seen these are qualities that, in Arendt's terms, define action more than they do a work of art: they are impermanent, leave no tangible traces and are activated through speech. This then may allow us to resituate the relationship between action and a work of art as less clearly defined, in the context of a shifting perspective upon what action means in contemporary practices.

In her famous essay on Arendt, Julia Kristeva (2001) highlights the central distinction of the Human Condition, that between the 'who' and the 'what'. To her, political action is essentially an actualization of the 'who'. Public space is the space where the 'who' – as a person defined by his/her actions – appears. The 'who' as revealed in action is also one of the issues in performative contemporary art, concerned as it is with giving appearance to a body that ceases to remain hidden as well as to the 'who' itself.

Kristeva exposes the contradiction inherent to Arendt's thinking: the biological and the body becomes unfit for participation in public appearance. To Arendt, 'what' someone is can be reduced to social appearance and biological attributes, while 'who' someone is resides in a separate being, one that appears to others but "remains hidden from the person himself". Even if hidden from one's self, the 'who' can appear to the other and be undisclosed to the memory of other people. And yet, the 'who' is revealed only in action and once this action "transcends the mere protective activity", it becomes a source of creativity. (Kristeva, 2001, p. 175)

Kristeva then suggests that because the 'who' knows that it is mortal and belongs to the spoken memory of multiple and conflicting opinions, it ceases to be a 'what' and attempts to transform both labor and work into an action, which is itself spoken, projected toward the past and the future, and shared with other people. The 'who' would then be, for Arendt, the only actor of the public sphere. As opposed to the 'who', the body itself never transcends nature and is deemed to remain attached to the sphere of privacy (as are women and slaves, who personify the body in labor). The body becomes for Arendt the paradigm of private property, hence apolitical. Kristeva argues that the Arendtian body is not simply apolitical, but also generic: Arendt considers both our inner psychic ground and our inner organs always to be the same, for since they do not appear, they are fundamentally apolitical.

Yet the body is also a way to express difference, and through its very manifestation as a difference (each body being intimately different and other), it is always political. Late contemporary art practices have explored and shown to what extent our body is a political issue and bearer of political meaning. In

Wodiczko's public projections, the body of the monument becomes and embraces the body of the speaker and through this interplay between body and architecture, the essentially political feature of the body is emphasized. In the Tijuana Projection, violence against women – and their female bodies – is brought into the public through the skin of a monument that becomes the skin of the person speaking of her traumatic experience as a woman, essentially different through the experience of her body. The strong political connotations of the body and its appearance in the public space, as shown by these works, thus questions Arendt's position in regards to the body and its political meaning.

Work of art *as* action

Five decades separate us from Arendt's influential text. In 1958, she could hardly have anticipated the present evolution of contemporary art that sometimes use action and speech as a medium for addressing issues such as the cultural and sexual Other, the way of being in public, political dominant discourses and critique of official culture.

Arendt sets strict boundaries between action and work as essentially different activities of the *vita activa*. Even though it occupies a privileged place in the realm of objects, the work of art never transcends its objecthood and worldliness. Yet the definition of a "work of art", one of the categories Arendt addresses, has changed considerably since the end of 1950s. Performance art, social art, new genre public art and site-specificity, including evolution from the site, have all been continuously challenging and testing the institutional definition of art.

Less objects than processes doomed to a short life, yet always involving the presence of others to see and listen, contemporary works such as those of Wodiczko and Holzer seem to challenge the boundaries between works of art and action and reveal a space of speech and appearance closer to action as defined by Arendt, action that constitutes public space. The examples of Wodiczko and Holzer illustrate a whole range of similar artistic practices that could further extend this discussion. Artists like Michael Asher, Hans Haacke, Christo and Jeanne-Claude, Joseph Kosuth, Daniel Buren and, to some extent, Richard Serra, use architecture, text, or both as medium, turning site-specificity into an ethnographic and socio-political interest.[57]

These practices help us understand Arendt's conception about action in relation to works of art. They propose a way of rereading Arendt that actualises

---

57 Hal Foster describes this as the psychoanalytic-anthropological turn (Foster 1996, pp.171-185).

her description of work and action through the lenses of these practices, one that allows for challenging and refining the distinction. At the same time, Arendt's influential theory of public space allows us deeper understanding of how contemporary artistic practices can engage and question the construction of such.

These works, characterized by temporality and relation to a political agenda that aims to make hidden conflicts or trauma visible, reposition the categories of action and work of art, within Arendt's own definition, and make them less clear-cut and more porous. This porosity could allow installing a "work of art" as a transitory category between action and work. This is arguably the proper place for the ephemeral art work offering visibility to the invisible through the action of the word and the power of the voice, giving appearance to the marginal, the deprived, the hidden, and the contradictory, challenging the condition of the Stranger.

References

ARENDT, H. (1999) *The Human Condition*. 2nd ed. Chicago: University of Chicago Press. (first edition 1958)
ARENDT, H. (1993) *Between Past and Future*. 3rd ed. Penguin Classics. (first edition 1961)
ARENDT, H. (1989) *Lectures on Kant's Political Philosophy*. 2nd ed. Chicago: University Of Chicago Press.
AUGÉ, M. (1992) *Non-lieux*. Paris: Seuil.
BAIRD, G. (1995) *The Space of Appearance*. Cambridge, Massachusetts: The MIT Press.
BENHABIB, S.(1990) Hannah Arendt and the Redemptive Power of Narrative. *Social Research*, 57(1), pp. 167–196.
CERTEAU, M. de, GIARD, L. and Mayol, P. (1990) *L'invention du quotidien, tome 1 : Arts de faire*. Paris: Gallimard.
CURTIS, K. (1999) *Our Sense of the Real: Aesthetic Experience and Arendtian Politics*. New York: Cornell University Press.
DEUTSCHE, R. (1996) *Evictions: Art and Spatial Politics*. Cambridge, Massachusetts: MIT Press.
FOSTER, H. (1996). *The Return of the Real*. Cambridge, Massachusetts: The MIT Press, pp. 171–185.

FRAMPTON, K. (1979) The Status of Man and the Status of His Objects: A Reading of the Human Condition, in ARENDT, H. (1979) *The Recovery of the Public World.* New York: St.Martin's Press.
FRY, K., (2001) The role of aesthetics in the politics of Hannah Arendt. *Philosophy Today*, 45(5), p. 46.
HABERMAS, J., (1990) *Strukturwandel der Öffentlichkeit: Untersuchungen zu einer Kategorie der bürgerlichen Gesellschaft.* 11th ed. Berlin: Suhrkamp Verlag (first German edition 1962)
HOLZER, J. (1998) *Jenny Holzer (Contemporary Artists).* London: Phaidon Press
JOSELIT, D. (1998) Voices, Bodies and Spaces: the Art of Jenny Holzer, in HOLZER, Jenny. *Jenny Holzer.* London: Phaidon Press.
KRISTEVA, J. (2001) *Hannah Arendt.* New York: Columbia University Press.
Krauss, R. (2010). *Perpetual Inventory.* Cambridge, Massachusetts: The MIT Press
LOUIS, E. (2006) Language in Visual Art, in DINKLA, S. (2006) *Jenny Holzer. Die Macht des Wortes./ I Can't Tell You: Xenon for Duisburg, the Power of Words.* Hatje Cantz Verlag.
RANCIERE, J. (2000) *Le Partage du sensible : Esthétique et politique.* Paris: La Fabrique.
VENTURI, R., Steven Izenour, and Denise Scott Brown (1977) *Learning from Las Vegas – Revised Edition: The Forgotten Symbolism of Architectural Form.* Cambridge, Massachusetts: The MIT Press.
VILLA, D. (1999) *Politics, Philosophy, Terror,* New Jersey: Princeton University Press.
VILLA, D. (2001) *The Cambridge Companion to Hannah Arendt.* 1st ed., Cambridge University Press
WODICZKO, K. and SICHÈRE, M.-A. (1995) *Art public, art critique: Textes, propos et documents.* Paris Musées.
WODICZKO, K. (1999) *Critical Vehicles: Writings, Projects, Interviews* Cambridge Massachusetts: The MIT Press.
WODICZCKO, K. (2001) An Interview With Krzysztof Wodiczko. Available from: http://www.art-omma.org/NEW/past_issues/theory/07.
WODICZKO, K. (2003) A Dean & DeLuca Coffee Shop Conversation with Krzysztof Wodiczko. *Perspecta 34* pp. 70–73.
WODICZCKO, K. (2004) *Making Critical Public Space.*agglutinations.com. Available from: http://www.agglutinations.com/archives/000035.html.
WODICZKO, K. (2005) *Interview & Videos | PBS Hiroshima Projection.* Available from: http://www.pbs.org/art21/artists/wodiczko/clip1.html

CHAPTER 6

Shaping Spaces of Shared Experience: Creative Practices and Temporal Communities

Jekaterina Lavrinec, Oksana Zaporozhets

Introduction

The paradox of current Urban Studies is a wide application of the concept of creativity to whole cities, but not to particular sites within (Landry, 2000; Florida, 2002; Evans, 2009). Although creative campaigns and actions obviously shape both urban and virtual scapes, fascinating and inspiring the public and turning thousands of Internet users into devoted followers of the creative groups who initiated them, the actions themselves rarely become an issue for analysis. Being brought to life, these creative practices as "a new combination of bodies (actions and passions, which are strung together...) and ... the verbal statement as result, as effect of the corporeal combination" (Lazzarato, 2003) make new urban scenarios possible. They enrich and reinvent urban everyday life by filling it with vividness and emotionality.

The study of creative practices is both an urban adventure and a theoretical challenge necessitating the revision of reflexive tools applied to urban settings and the very notion of the urban itself. Understanding the city as a changing configuration of practices is the basic attitude of the micro-perspective developed by de Certeau (de Certeau, 1988) and Lefebvre (Lefebvre, 1991). Surprisingly, the idea of practices both shaping urban space and being shaped by it becomes the statement, which has too long been taken for granted in Urban Studies instead of being problematised and developed. We consider creative practices as the means of reconfiguration for the cityscape and the intellectual-scape of Urban Studies. Giving birth to "urban enthusiasts" who reinterpret and reinvent the city combining the means of reflection with the means of knowledge, these creative practices add a new figure to the list of urban characters such as flâneur (Benjamin, 1968; Benjamin, 2006) and drifter (Debord, 2006a).

Creative practices increase the diversity of urban scenarios attached to particular places and the intensity of urban emotions. Involving citizens in creative events, they change the routine urban choreographies and increase the significance of bodily contact, synchronized movements, and simultaneous emotions. Thus, the urban places where these practices are set become "spaces of shared experience" formed by mobile situations or prolonged rituals.

Applying micro-optics: co-creating urban space

Describing the art of "making do", de Certeau points out the practices constituting everyday life or the "ways of using" (de Certeau, 1988, p. 35). These invisible ways of dealing with things, spaces, and time create the essence of the everyday life revealed by the researcher: "[t]he practices of consumption are the ghosts of the society that carries their name" (de Certeau, 1988, p. 35). de Certeau concentrates on these ways of using and almost avoids the conceptualization of the actors, preferring to describe them through the practices they are involved in (for instance, the opposition of strategies and tactics) and their more or less materialized footprints.

Applying micro-optics, which actualizes the state of being in the city, we will bring the actor to the urban scene[58]. The "actor's comeback" is a part of the oligopticon vision[59] (Amin & Thrift, 2002; Latour, 2005) allowing us to look at urban life from different sites "to see little, but to see it well" (Latour, 2005, p. 181). The "humanization" of these practices makes evident the omissions in de Certeau and Lefebvre's theoretizations. Their concentration on practices implicitly inscribes an individual or an aggregate actor such as the "authorities", "ordinary people", etc. Changing the scale of consideration, we will here focus on the co-actions constituting everyday life and shaping urban spaces. The theoretical frames of co-action developed by the Symbolic Interactionism (see: Goffman, 1961) or Situationism (Debord, 2006b) are undeniably insightful but at the same time quite restrictive in their understanding of urban life, stressing either an abstract scenario of social interactions or their political significance.

We begin our analysis of urban co-actions by acknowledging a tendency of mutation of *researcher* into *reflexive activist*. This figure is quite similar to what Knight calls the "outsider artist", keeping in mind activists whose aims and actions lay "outside of the art world's conventions and constrictions" (Knight 2009, p. 115). The reflexive activist is conscious about urban problems and tendencies and reacts to them by initiating actions in urban space, which are addressed to a wide audience. S/he considers *the urban art interventions* and *creative actions* to be the inseparable "means of knowledge and the means of action"[60] (Chtcheglov 2006). The urban interventions and creative actions are

---

58 This study comprises research findings from the project 'Graffiti and street art in the cultural space of big cities" carried out within The National Research University Higher School of Economics' Academic Fund Program in 2012, grant No 12-05-0002

59 Here we understand oligopticon as "series of partial orders, localized totalities, with their ability to gaze in some directions not others" (Amin, Thrift, 2002, p. 92)

60 Ivan Chtcheglov in Situationist "Formulary for a New Urbanism" develops the idea of reflexive action as a key principal of new architecture, which is supposed to be "a

comprehended as a tool for re-inventing and revitalising urban settings while initiating intensive interaction and cooperation between citizens. Considering interventions and creative campaigns as the process of co-action, we admit that at some point the interaction based on the active involvement of urban dwellers dissolves the difference between the initiator of the event and its participants. Shared space and actions shaping the common experience and intense emotional atmosphere of the event turn the active actors into the new group, which we call *"the urban enthusiasts"*. The notion of urban enthusiasts does not eliminate the other positions and distinctions kept alive during the interventions, rather it stresses the new form of solidarity and permanent conversion of the roles of the initiators of and participants in the creative campaigns.

This space of co-action filled with co-presence, communication, and emotions becomes "the space of shared experience"[61]. We develop this term to explicate the micro-perspective on public place and demonstrate its heuristic potential. The term "space of shared experience" is not seen as an alternative or a substitute of "public place". Instead, it opens a new layer – the layer of instant actions that reinterpret the scenario of the space, creating its own story, one that is evidently missed in numerous approaches conceptualizing public space mainly as a part of a general social landscape (modern city (Sennett, 1977), symbolic production (Zukin, 1995), or group interaction (Carmona, 2010), etc.) .

A position toward urban environment, which we define as enthusiastic and which implies active participation in shaping urban space, is close to the situationists' idea of initiating "emotionally moving situations" in the city. Reconstructing the concept of "situations", which was the essential part of the Situationist International's program, Simon Sadler points out that "the

---

means of modifying present conceptions of time and space... Architectural complexes will be modifiable. Their appearance will change totally or partially in accordance with the will of their inhabitants" (Chtcheglov, 2006).

61 Using the category "space of shared experience", we stress the both the process of sharing and the new scenarios produced while sharing the urban space. The suggested category has some common ideas with the notion of "shared space" developed by Monderman in civic engineering, although it is evidently more general. Monderman and his team describe the shared space as the encapsulation of "a new philosophy and set of principles for the design, management and maintenance of streets and public spaces, based on the integration of traffic with other forms of human activity. The most recognizable characteristic of shared space is the absence of conventional traffic signals, signs, road markings, humps and barriers – all the clutter essential to the highway. The driver in shared space becomes an integral part of the social and cultural context, and behaviour (such as speed) is controlled by everyday norms of behavior". See: Project for Public Spaces http://www.pps.org/shared-space/ . Retrieved 25. 07. 2012.

constructed situation would clearly be some sort of performance, one that would treat all space as performance space and all people as performers" and each of the situations "would provide a décor and ambiance of such power that it would stimulate new sorts of behaviour, a glimpse into an improved future social life based upon human encounter and play" (Sadler 1999, p. 105). Recognizing the importance of emotions in urban experience, situationists developed a vision of urban space that is open to changes brought by citizens: e.g. Debord sees the potential of architecture in "emotionally moving situations, rather than emotionally moving forms" and believes, that "experiments conducted with this material will lead to new, as yet unknown forms" (Debord 2006 b).

An important insight of the situationist movement was the idea that spatial structures produce certain types of behaviour as well as an emotional experience (a key element of psychogeographical mapping) and that said structures are able to "activate" citizens and encourage them to take part in the construction of their urban surroundings. From this perspective, creative actions and urban art interventions constitute a quite productive method of articulating urban experience and rearranging the emotional landscape of the city. By disturbing the usual everyday rhythms and trajectories, urban interventions establish a reflexive distance from the routine choreography of the place and propose alternative scenarios of behaviour in public space. Therefore urban art interventions and urban games can be considered as a tool for the creative reconceptualization of spatial structures and social order, embedded in urban space.

Creative practices and the space of shared experience

What we call creative practices is an active reinterpretation of urban space, which is produced by urban enthusiasts. Creative campaigns (flash mobs, urban games, and various other public actions) and urban art interventions provide a possibility of publicly-shared emotional and bodily experience and establish momentary citizen solidarities. While art interventions reinterpret spatial structures using the potential of the place for further reorganizations, creative actions reinterpret routine scenarios embedded in various types of urban spaces (airports, squares, train stations, shopping malls, bridges, etc.). But in fact, these two vectors of reinterpretation of urban space are interconvertible, as a long-term art intervention in an urban space can initiate alternative scenarios of behaviour in a particular place, and some creative actions become urban ritual (a repetitive symbolic action, connected with a certain urban place or element and which reinterprets not only spatial, but also temporal structures of urban places).

As an example of an art intervention that radically changes the routine choreography of a certain place, we may consider "Miracle tile", an installation created by Lithuanian artist Gitenis Umbrasas. A tile with a word "Miracle" inlayed in it was installed on the Cathedral Square in Vilnius (Lithuania). As it differs from the other tiles, it naturally caught attention of the passers-by and inspired them to search for an interpretation of this element, a tile. The tile became an object of active bodily interpretation: while making a wish, citizens began stepping on the tile. This developed into some people turning around on the tile and others jumping on it. The ritual of turning around seemed to 'catch on' and has expanded among citizens over the several months that followed the tile's installation.

As a new urban element, the tile proposed a certain urban choreography: it turned into an off-beat on the way of passers-by who cross the square. After the tile was installed, passers-by received a place to stop and to perform simple movements when crossing the space. The instructions on how to make a wish have spread among citizens spontaneously and now are usually included in urban narratives for tourists. It is no wonder that the "Miracle tile" became a point of attraction for groups of tourists but, what is more important, this small urban element became a place of shared emotions and the articulated intimate experience of making a wish.

Art interventions, as well as performances and flash mobs, are a heuristic tool that reveals the interconnections between spatial structures and everyday practices by creating points of attractions (or spatial obstacles) and disturbing everyday routes and scenarios. In this sense, art interventions sync quite closely with the idea of situationists, that urban routine must be "disturbed" in order to produce conditions for the re-examination of everyday experience. Interventions propose micro-practices, which differ from routine choreography: e.g. stopping in a crowded place to throw back a head, rubbing the hand of a street sculpture, leaning over some object to scrutinize it, jumping over some obstacle, etc. Those art interventions that affect the bodily experience of passers-by in a playful way or leave a possibility for establishing a bodily connection saturate a place with vivid emotions and usually become a point of attraction for locals and travellers, which after some time turn into an official sight.

Creative events (dancing flash mobs, hugging campaigns or a campaign of a salutation of passengers who alight from public transport) reinterpret actively routine scenarios and reinvent scenarios of social interaction by introducing alternative models of behaviour to a certain type of place (railway stations, bus stops, lecture rooms, big shops). Many alternative scenarios, invented by urban enthusiasts, spread across the world and can be repeated in every similar urban place[62]. Being a type of "open-ended medium with endless variables" (Goldberg, 2001), creative events broaden the scope of urban scenarios attached to particular places and to urban settings in general. Shaking the very idea of urban alienation and the "right to be left alone" (Tonkiss, 2005), they are not targeted to explode or completely substitute for urban conventions but rather to awake citizens from somnambular walks and make them more sensitive to the variety of scenarios potentially suggested by urban places and city in general. Despite this opportunity, the walkers still have a chance to skip the event, to pass it by and restore distance and emotional "neutrality" as well as to get a feeling of security by hiding themselves in the "safe bubble" of urban conventions.

Creative campaigns (be it hugging, dancing, or mass reading in public) imply certain choreography of participants, different from their usual behaviour and as a result, provoking vivid emotions. Such campaigns are aimed at producing a certain emotional effect, e.g. at surprising and sharing the joy with casual passers-by: as announced at the official site of the "Improv Everywhere" collective, they cause "scenes of chaos and joy in public places" (Improv Everywhere 2001[63]). Initiating flash mobs and urban games is also a temporal solution for the problem of hunger for emotions and bodily contact. According to the initiator of the worldwide "Free Hugs Campaign", the inspiration for this campaign was the experience of loneliness he had when arriving in Sydney after being absent for a long period of time:

"Standing there in the arrivals terminal, watching other passengers meeting their waiting friends and family, with open arms and smiling faces, hugging and laughing together, I wanted someone out there to be waiting for me. To be happy to see me. To smile at me. To hug me" *(Free Hugs Campaign 2011[64])*.

---

62 Some of the actions can spread spontaneously, such as the so-called love padlocks. Padlocks with the engraved names of couples get affixed to fences and railings of bridges; the keys are thrown away after the padlock is locked. This ritual saturates the place with romantic feelings. Even those who have never taken part in this ritual and have never before heard about it recognize the structure of symbolic action. And maybe it is partly because of its recognizable structure that this urban ritual has expanded into many cities across the world. It takes only a few weeks for several padlocks to appear on the railings of a bridge; after a couple of months more no free place will be left there.
63 See http://improveverywhere.com/.
64 See http://www.freehugscampaign.org/.

In many creative events passers-by are invited to be the participants of the action. Moreover, the initiators of such campaigns as "Free Hugs" or the series of dancing or singing performances arranged by "Improv Everywhere" use the potential of urban spaces for establishing contact with passers-by. It is the same scenography of flash mobs and performances that reconceptualizes the relations between the performer and the viewer, and reinvents the idea of proximity. As a rule, participants of flash mobs and urban games "dissolve" in the crowd before the action starts. Even in cases where passers-by remain in the position of viewers, still the usual distance is being overcome, as participants of flash mobs or performances appear to be a random person next to the viewer. Thus, in a public musical arranged by the T-Mobile creative group at Heathrow airport ("Welcome Back", 2010), singers are the part of the crowd. Some of them have mixed into the group waiting for arriving persons and some emerge from the arrival gates. In this case, only singers were equipped with microphones and others were not encouraged to sing; instead, some of the viewers became addressees of welcome songs. The common experience of arriving, departing and welcoming became a ground for building a new experience of the same place (airport), which turned into a playground for sharing joy and surprise. The reaction of the viewers is an essential part of the event and has become an important part of any records of the event, as it communicates the same atmosphere of the event and its emotional saturation.

By initiating and participating in urban events, which set up an alternative model of behaviour, and sometimes reshape the usual temporal model by mixing usual scenarios of leisure and work, citizens establish new solidarities. Being an alternative to the monotony of urban everyday life, creative campaigns provoke intense emotional reactions from citizens. Appealing to the reaction of the passers-by, they give a way to a variety of feelings and emotions based on surprise. Surprise becomes an emotional trigger that leads to the changing emotions (suspicion, interest, fascination, etc.) that saturate subsequent communications and actions. This rapidly changing variety of feelings contrasts with the typical emotional cityscapes, which usually serve to actualize the "stable" emotions attached to places, such as boredom, grief, joy or fear.

These creative practices do not simply break the flow of everyday life. They foster and prolong the state of uncertainty and the spontaneity of the situation as grounds for social creativity. The uncertainty of the situation provokes actors to coordinate their reactions "here and now", adjusting to the moment and to the actions of the others and, at some point, also to avoid the "usual" schemes, which are inappropriate or ineffective. So, the creative campaigns foster not only the very fact of interaction, they also develop communication skills and valorize the state of creativity and mutual adjustment.

This is the case when the logic of social exchange (as a basis for solidarity building) substitutes, although temporarily, for the mainstream scenario of city spaces. One of the implicit schemes undermined by the creative practices is the commoditization of urban life – the essential characteristic of the modern city (Simmel, 1998; Lefebvre, 2008, etc.). Doing something not for money is the motto unifying the wide range of urban activists (from guerila gardeners to street-artists, from performers to interventionists or different type of "bombers" (Reynolds, 2009)) and has been enthusiastically supported by citizens. Emphasizing the significance of social skills, the logic of enthusiasts and wonderers becomes the basic stepping block for new solidarities. That is why the issue of "being sold out" as undermining the very idea of social cooperation (the core value of these associations) has become a frontier issue for such communities. The social exchange based on the bodily co-presence, communication, and shared emotions devaluate the urban convention of being "socially distant yet physically close" (Wirth, 1938). Physical closeness supplemented with communication and orientation to the other gives the actors of the creative campaigns a chance to manage the social distance, to balance the individual and collective efforts.

The scandalous action "Dick Captured by the FSB" of the Voina group, which took place in Saint-Petersburg in 2010[65], became the local top news story at the time, filling the city with rumors, guesses, and talk. Here is an example of one blogger's comments on the citizens' reaction: "Everybody talked about it in Saint-Petersburg... everybody saw it. Even those who didn't see it, they definitely knew about it"[66]. The event temporarily changed the social and emotional scapes of the city. The emotional richness and energy of creative actions and their adventurous inclinations are key moments giving a birth to new urban solidarities. The "unusual" intensity of emotions adds a special importance to the common experience, signifying it as a different compared to the routine "ethics of indifference" (Tonkiss, 2005) or "mutual neutrality" (Sennet, 2010). This intensity and variety makes the emotional cityscape more diverse and vivid.

---

65 "Dick captured by the FSB" the action of Voina group including the drawing of the penis on Liteiny Bridge in 23 seconds. It is an erecting bridge with the FSB building situated in its surroundings.
66 See: http://plucer.livejournal.com/262707.html

Virtual co-being: media platforms and temporal solidarities

Shared emotions and actions produce short-term as well as long-term solidarities influencing the contacts of citizens and widening the audience of an event. The more emotionally intensive an event is, the more it breaks the routine of a place, the more it impresses citizens and intensifies the social nets of communication. The role of media here can be crucial: the archive of such events (writings, photos and videos accessible in the Internet) is also a chronicle of the temporal solidarity, the legitimation and the evidence of its fluid or permanent existence. Internet platforms (such as flickr, youtube, facebook, twitter, and blogs) provide conditions for sharing the experience of taking part in the events, attracting more potential participants and transcending the local context of the creative action[67]. Recording media and interactive media (especially, mobile phones and Internet) are important tools in constituting urban creative events, enhancing them with their technological opportunities.

Recording the event (or the reaction of passers-by toward an intervention) and sharing pictures and videos afterwards has become a form of participation. As soon as participation of the videographers became an essential ingredient of creative campaigns in urban space, the choice of place for the action began to take into consideration the spatial potential of a place to provide several surveying points. For the documentation of mass actions, such as dancing flash mobs, it is important to record a whole group of participants and the reactions they cause. The place itself usually is interpreted more as a scene for action than as a particular site with its specific traits. Often for flash mobs and creative campaigns (such as mass singing in squares), sensitivity towards the spatial structures of a place does not imply attention to the identity of the place itself. In these cases, information on the location of the event is provided by commentaries, left by photographers or operators. The action itself, the effect it causes, the emotions of participants and the reactions of viewers is the main object of documentation for many creative campaigns. And records, shared on the Internet, provide valuable instructions on how to arrange similar events and encourage enthusiasts to repeat the experience in other cities and countries. In the case of creative campaigns, which are aimed at revitalizing concrete places in a city, place identity becomes a topic of concern for photographers[68]. Creative

---

67 One of many examples is the initiative "Dispatchwork", the idea of which was to reconstruct urban objects using colorful construction sets and became popular in many countries. See: Dispatchwork.info (www.dispatchwork.info).

68 For example, the creative campaign "Bubble your city!" ("Burbuliatorius"), which was initiated in Lithuanian cities during 2009-2011 and which brought citizens to

reinterpretation of spatial structures of particular places become a main topic of documentaries, devoted to urban art interventions.

Shared on youtube, flickr, and at social networking services, these records become a point of attraction not only for participants of the event and thus widen the auditory scope of the event. The interactive media contribute to creating an augmented public space for a wider audience by making the event more visible and activating communication and debate, leaving the meanings of the event open-ended and ready for reinterpretations. Being a "witness of the event" and acting as the virtual billboards announcing the event, platforms of interaction and instant communications, free accessed archives, and media "rather enhance than annihilate" the urban life (Aurigi, De Cindio, 2008). For instance, sharing the information and photos of the event is the next step in the cooperation and community building inspired by the actual event, now realized in virtual space.

This media communication explicates and reinforces convertibility as an important trait of creative gatherings. Thus, the convertibility of the roles from viewer/follower to participant and vice versa increases the openness and attraction of the event. Just as a passerby might be occasionally turned into a participant in creative actions, the Internet follower of a group might become the initiator or the "agent" of the next performance (as it happens with Improv Everywhere) or simply join the group by sharing emotions inspired by the action. Some virtual communities, based on the shared interest of taking part in creative actions, remain active even during periods when no events are being initiated by the creative group (for example, during cold winter time). It is also quite possible that on the basis of a community that gathered around a special kind of event, some new ideas and alternative sub-groups emerge[69].

---

deactivated squares for periodical playful activities, became a series of photos of representational places of Lithuanian cities, which are now used gladly by tourism centres. The periodic urban event, which brings citizens of all ages to the center of the city transformed into a type of photography symposium, where photographers exchanged their professional experience and searched for new visual solutions to documenting the event. The event provided a ground for the emerging of professional communities of city photographers.

69 For instance, before-mentioned "Bubble your City" event ("BC"), organized by "Laimikis.Lt" group in Lithuania, inspired the emergence of "Sofa evenings" activities in the city of Kaunas (2010-2011). "Sofa evenings" was the initiative of Kaunas citizens, who created a home-like space at the central alley of the city, by the place where "BC" is being arranged, and use it for concerts and playing board games. This initiative prolonged "Bubble your City!" gatherings, proposing alternative formats of leisure in the city and leading to the periodic reviving of the area.

Another type of convertibility is the virtual/real transformation. Media can in fact facilitate the de-virtualization of social communication by becoming "a catalyst for gathering and community activities" (Uricchio, 2009) or by means of the "echo effect" (Auge, 1995) – reproductions of events in other urban surroundings. They also work as "virtual stabilizers" by prolonging the effect of an event in media space, keeping alive the solidarities that emerged and stimulating new ones. For the participants of the event, its prolongation in the media (especially in the form of internet communities) provides a perspective for further development.

Beyond all of this, various media also provide "the frame of orientation" articulating, debating and legitimating new meanings of the places and new urban scenarios. For example, the fact that the mentioned action of the Voina group turned Liteiny Bridge into an urban attraction with a new meaning attached to it is ironically reflected in the title of another *youtube* video, "Liteiny Bridge. Not about dick!" Although the previous meanings seemed to be unquestionable, stable, and shared by the Saint-Petersburg citizens, the new one successfully competes with them, turning the urban symbolic landscape into the palimpsest of senses.

Creative events enforce the role of the media as the tool of "emotional management", giving the viewers the opportunity to run the communication and appropriate the event (for instance, increasing its emotional intensity through banal "likes" in social networks or comments). The prolongation of the creative event in the media somewhat "neutralizes" its momentary-ness and spontaneity located in urban spaces, contrasting with the logic of stability inscribed in urban materiality and interactions and valorizing the ephemerality of the actual event.

Conclusion

Numerous creative practices being suddenly brought to life have impressively reconfigured urban and virtual scapes over the last decade. Creative actions widen the experience of the city, attaching new values and meanings to certain places. Urban creative practices re-interpret the spatial and temporal urban structures and regulations affecting the everyday bodily and emotional experience of urban dwellers. Producing new forms of solidarities and co-actions, they challenge the very notion of the urban as well as the reflexive tools and concepts applied towards understanding urban life.

The choreographies of creative campaigns involve citizens in new, playful, spontaneous, and emotionally intensive co-actions, undermining the distinction between initiators and viewers. The experience of communication based on spatial and emotional proximity forms new social skills, presents an alternative to distant, neutral, and prescribed urban communications, and creates the "space of shared experience". Thus, creative campaigns become an improvised educational platform developing and introducing new social scenarios and training communicational skills. Being media-ized and presented on the web, scenarios of creative actions spread worldwide. The web-platforms are turned into the virtual site of community support and building, preserving established contacts and facilitating new ones.

We hope that this concentration on the "spaces of shared experience" will provide material for questioning the social, spatial, and many other structures shaping the cityscape, as well as the established notions such as "public spaces" that make up the landscape of Urban Studies.

References

AMIN, A. and THRIFT, N. (2002) *Cities: Reimagining the Urban.* Cambridge: Polity Press.

AUGE, M. (1995) *Non Places. The Introduction to Anthropology of Supermodernity.* London, New York: Verso Books.

AURIGI, A., De CINDIO, F. (2008) *AugmentedUrbanSpaces. Articulating the Physical and Electronic City.* London: Ashgate Publishing.

BENJAMIN, W. (1968) Paris – the Capital of Nineteenth Century. *New Left Review I/48,* pp. 77–88. Available from: http://www.newleftreview.org/?page=article&view=134

BENJAMIN, W. (2006) Hashish in Marseilles. In: BENJAMIN W. *On Hashish.* Harvard: Belknap Press of Harvard University Press.

CARMONA, M. (2010) Contemporary Public Space: Critique and Classification, Part One: Critique. *Journal of Urban Design.* 15: 1, pp. 123–148.

CHTCHEGLOV, I. (2006) (1958).Formulary for a New Urbanism. In: KNABB, K. (Ed.). (2006) *Situationist International Anthology.* Berkeley: Bureau of Public Secrets. Available from: http://www.bopsecrets.org/SI/Chtcheglov.htm

DEBORD, G. 2006 a (1958).Theory of the Dérive.IN KNABB, K. (Ed.). *Situationist International Anthology.* Berkeley: Bureau of Public Secrets. Available from: http://www.bopsecrets.org/SI/2.derive.htm

DEBORD, G. 2006 b (1957). Report on Constructing Situations. In: KNABB, K. (Ed.). *Situationist International Anthology.* Berkeley: Bureau of Public Secrets. Available from: http://www.bopsecrets.org/SI/report.htm

DE CERTEAU, M. (1988*) The Practice of Everyday Life.* Berkley: University of California Press.

EVANS, G. (2009) Creative Cities, Creative Spaces and Urban Policy. *Urban Studies. Vol. 46, 5–6: pp. 1003–1040.*

FLORIDA, R. (2002). *The rise of the creative class—and how it is transforming leisure, community and everyday life.* New York: Basic Books

GOFFMAN, E. (1961) *Encounters. Two Studies in the Sociology of Interaction.* London: Macmillan Pub Co.

GOLDBERG, R.L. (2001) *Performance Art: From Futurism to Present.* London: Thames & Hudson.

KNIGHT, Ch. (2009) *Public Art: Theory, Practice and Populism.* Blackwell Publishing

LANDRY, Ch. (2000) *The Creative City: A Toolkit for Urban Innovators.* London: Earthscan Publication.
LATOUR, B. (2005) *Reassembling the Social. An Introduction to Actor-Network-Theory.* Oxford: Oxford University Press.
LAZZARATO, M. (2003) *Struggle, Event, Media.* Available from: http://republicart.net/disc/representations/lazzarato01_en.htm
LEFEBVRE, H. (1991) *The Production of Space.* Oxford: Blackwell Publishing.
LEFEBVRE, H. (2008) The Right to the City. In: LEFEBVRE, H. (auth.), KOFMAN, E. and LEBAS, E. (Eds.) (2008) *Writings on Cities.* London: Wiley-Blackwell.
REYNOLDS, R. (2009) *On Guerrilla Gardening.* London: Bloomsbury Publishing
SADLER, S. (1999) *The Situationist City.* London: The MIT Press.
SIMMEL, G. (1998) The Metropolis and Mental Life. In: SIMMEL, G. (auth.) FRISBY, D., FEATHERSTONE, M. (eds.) *Simmel on Culture.* Selected Writings. London: Sage Publications.
SENNETT, R. (1977) *The Fall of Public Man.* Cambridge: Cambridge University Press.
SENNETT, R. (2010) The Public Realm. In: BRIDGE G., WATSON, S. (Eds.) *The Blackwell City Reader.* Oxford: Blackwell Publishing.
TONKISS, F. (2005*) Space, the city and social theory.* Oxford: Polity Press.
URICCHIO, W. (2009) The Future of the Medium Once Known as Television. In: SNICKARS, P. and VONDERAU, P. (Eds.) (2009) *The YouTube Reader.* Stockholm: National Library of Sweden.
WIRTH, L. (1938) Urbanism as a Way of Life. *American Journal of Sociology.* Vol. 44, No. 1, pp. 1–24.
ZUKIN, S. (1995) *The Cultures of Cities.* Oxford: Blackwell Publishing.
PROJECTS AND ACTIONS
Project for Public Spaces: http://www.pps.org/shared-space/. Retrieved 25.07.2012
Improv Everywhere (the New York City based prank collective) http://improveverywhere.com/. Retrieved 25.07.2012
Free Hugs Campaign http://www.freehugscampaign.org/ . Retrieved 25.07.2012
T-Mobile flashmob at Heathrow airport (October, 27, 2010). http://www.guardian.co.uk/media/2010/oct/27/t-mobile-flashmob-ad. Retrieved 25.07.2012

Notes on Contributors

Clara Fohrbeck, is a Master Student in Sociocultural Studies at the European University Viadrina in Frankfurt Oder. At the moment she researches about international creative workers' new forms of working and living in Berlin for her Master's Thesis. Her research interests include a sociological perspective of new forms of employment and mobility, urban sociology connected to topics of aestheticization and the discourse surrounding creativity as well as gender-related topics.

Celia Ghyka, architect, PhD, teaches theory of architecture at the "Ion Mincu" University of Architecture and Urban Planning, Bucharest. Her interests refer to public space, memory, art and psychoanalysis, contemporary art and architecture, post-communism, themes that she has approached in several writings published in Romanian academic journals and books, as well as a guest lecturer in postgraduate joint courses at Ecole Nationale d'Architecture de Paris la Villette & Université de Paris VIII or Université de Luxembourg. Dr. Celia was a Getty-New Europe College fellow in 2002 and 2011.

Jekaterina Lavrinec, Ph.D., is a Vilnius based researcher in the fields of urban and media studies. She is a participant of Critical Urbanism Lab and a co-founder of NGO "Laimikis.lt", which develops scenarios for revitalization of public spaces and initiates cultural events. As an associate professor of the Department of Creative Entrepreneurship and Communication (Vilnius Gediminas Technical University) she teaches courses on urban culture and media studies.

Giulio Mattioli is a Ph.D Candidate in European Urban Studies (URBEUR). Since 2008 he has been collaborating with the Department of Sociology and Social Research of the University of Milano-Bicocca in research projects focused on urban mobility. His research interests include sustainable transport, car use, mobility & public space, accessibility & transport exclusion, and the sociological approach to climate change.

Oleg Pachenkov is a leading researcher at a state-independent research institute, the Centre for Independent Social Research (CISR, St.-Petersburg, Russia), and a director of the Centre for Applied Research (CAR) at the European University at St.-Petersburg. He received his doctorate degree (Candidate of Science in sociology) from the sociology department of St.-Petersburg State University in 2009. Currently he is working mainly in the fields of urban studies, applied urbanistic and interdisciplinary projects, binding social scientists together with architects, designers, urban planners and artists working with urban environment, street and public art. Dr. Pachenkov has published more than 30 articles in Russian, English and German addressing the issues of migration and ethnic entrepreneurship, informal economy, street vending and flea markets as phenomena of urban culture.

Laura Panait holds a PhD in Anthropology at the European Studies Faculty, Cluj. She is engaged both in research and in practice of public space and art interventions in Romania, working constantly in interdisciplinary projects with architects, artists and sociologists. She is also a member of the Paint Brush Factory. Now she is currently investigating the creative community in relation to the actual protests started in Romania at the beginning of 2012.

Tobias Scheidegger, PhD Student, teaches and researches at the Institute for Popular Culture Studies, University of Zurich, Switzerland. In his 2009 published master's thesis „Flanieren in ArCAADia" he analysed the culture of digitally rendered architectural visualisations, by focusing on their impact on the contemporary production of urban space. Furthermore he wrote several smaller articles about hegemonic design of urban public space and on the rhetoric of urban planning.

Lilia Voronkova is a social anthropologist, photographer, and curator. She has been working in the state independent research institute the Centre for Independent Social Research (CISR, St.-Petersburg, Russia) since 2003 as a researcher, and since 2011 as a coordinator of trans-disciplinary art-social science projects. Lilia facilitates the development of diverse forms of collaborative work between artists and scientists such as presentation of research projects to the public in the form of exhibitions, publishing catalogs, organizing art-science seminars and collaborative research projects. She has realized several art-science projects in the form of seminars, visual presentations and exhibitions.

Oksana Zaporozhets, Ph.D.(Candidate of Science in sociology), is currently a Moscow-based researcher in the field of Urban Studies. She is a leading research fellow at the Poletayev Institute for Theoretical and Historical Studies in the Humanities, National Research University Higher School of Economics (Moscow) teaching courses on Urban Studies and World Cities. She is a participant of the Laboratory of Critical Urbanism and a visiting lecturer at European Humanities University (Vilnius, Lithuania) giving a course in Urban Studies with Jekaterina Lavrinec.

www.ingramcontent.com/pod-product-compliance
Ingram Content Group UK Ltd.
Pitfield, Milton Keynes, MK11 3LW, UK
UKHW022212230426
12048UKWH00016BA/799